IP APPLICATIONS
WITH ATM

McGRAW-HILL SERIES ON COMPUTER COMMUNICATIONS (SELECTED TITLES)

IP Applications
with ATM

John Amoss
Daniel Minoli

McGraw-Hill
New York · San Francisco · Washington, D.C.
Auckland · Bogotá · Caracas · Lisbon · London
Madrid · Mexico City · Milan · Montreal · New Delhi
San Juan · Singapore · Sydney · Tokyo · Toronto

Library of Congress Cataloging-in-Publication Data

Amoss, John, date.
    IP applications with ATM / John Amoss, Daniel Minoli.
        p.   cm.—(McGraw-Hill series on computer communications)
    Includes index.
    ISBN 0-07-042312-1
    1. Asynchronous transfer mode.   2. Internet (Computer network)
3. Computer network protocols.   I. Minoli, Daniel.   II. Title.
III. Series.
TK5105.35.A46   1998   98-13201
004.6'6—dc21   CIP

McGraw-Hill

A Division of The McGraw·Hill Companies

1 2 3 4 5 6 7 8 9 0   DOC/DOC   9 0 3 2 1 0 9 8

ISBN 0-07-042312-1

*The sponsoring editor for this book was Steven M. Elliot, the editing supervisor was
Stephen M. Smith, and the production supervisor was Sherri Souffrance. It was set in
Vendome ICG by Michele Bettermann of McGraw-Hill's Hightstown, N.J., Professional
Book Group composition unit.*

Printed and bound by R. R. Donnelley & Sons Company.

McGraw-Hill books are available at special quantity discounts to use as premi-
ums and sales promotions, or for use in corporate training programs. For more
information, please write to the Director of Special Sales, McGraw-Hill, 11 West
19th Street, New York, NY 10011. Or contact your local bookstore.

 This book is printed on recycled, acid-free paper containing
a minimum of **50%** recycled, de-inked fiber.

CONTENTS

Contents

Contents

Contents

PREFACE

Overview

Asynchronous transfer mode (ATM) is a high-bandwidth, low-delay switching and multiplexing technology now available to corporate planners. It is being deployed in support of networks that carry an organization's traditional data and LAN traffic, video flows, image information, and even voice. In addition, nearly all major carriers in the United States (interexchange carriers, competitive local exchange carriers, and incumbent local exchange carriers) are now deploying or plan to deploy some form of ATM-based backbone. For example, ATM switches are being defined as platforms supporting services such as frame relay.

ATM, its services, and its promise are all heavily dependent on ATM's switching capabilities. ATM supports line speeds of 155 Mbps and 622 Mbps now, and will evolve to as high as 10 Gbps in the future. Today's switches support a combined throughput in the tens of gigabits per second range. To provide ATM-based services, including cell relay service, Internet Protocol (IP) traffic transport, video distribution, and voice, the switches that need to be deployed in public networks as well as in larger private corporate networks must handle throughput in the terabit per second range. New switching architectures are required to handle these traffic streams. Trunks terminating on such a switch will be fiber-based and may well reach into the 2-to-10-Gbps range (compared to the 45-, 155-, or 622-Mbps of today). Subscriber lines will also be fiber-based, and will operate in the 155-to-600-Mbps range; feeder systems for the loop plant will be in the 2.4-Gbps range. There are also stringent performance requirements on a switch where the input ports and trunks operate in the gigabit per second range.

Switching products utilized to support these services fall into three categories: core network switches, edge network switches, and user's (premises) switches, with a debate now under way vis-à-vis the migration of switching closer to the user, that is, away from the core switch and in favor of edge nodes (also known as service nodes). Vendors include traditional telco central office suppliers as well as a host of new vendors that are supplying the corporate community in terms of premises switches and hubs for private networks.

xiii

Purpose of this Book

This book provides essential information to assist corporate and carrier planners in making ATM a reality for their organizations. Users, equipment suppliers, and carriers need to understand ATM switching technology, economics, design considerations, and available carrier services in order to effectively deploy ATM and its capabilities in support of new, bandwidth-intensive applications such as digital video, desktop video conferencing, multimedia, imaging, and distance learning.

This book addresses two major areas. First, Chaps. 1 to 6 give a basic understanding of the technology and kinds of vendor products available today.

- Chapter 1 provides an overview of the major ATM opportunities in enterprise and carrier network applications. It discusses the current major industry drivers and barriers for ATM and presents the steps needed to be taken to ensure its successful deployment.

- Chapter 2 gives a detailed review of the technical aspects of the ATM reference architecture. This chapter presents a thorough review of the principles involved at the Physical Layer, with an emphasis on SONET technology; the ATM Layer, with its Quality of Service guarantees; and the ATM Adaptation Layer, which adapts the services provided by ATM to user requirements. This chapter also describes the ATM information flows in the user and control planes that have to be supported across an ATM network.

- Chapter 3 provides an up-to-date description of the on-going standards work that is defining various aspects of ATM technology. Emphasis is given in this chapter to the work of the ATM Forum.

- Chapter 4 looks at the architecture of an ATM switching system from a functional view, and describes key functions such as traffic and congestion management, routing, and connection management.

- Chapters 5 and 6 provide examples of commercial ATM switches. Chapter 5 describes ECI Telematics' 1E6 multiservice ATM switch architecture, and illustrates the various concepts discussed thus far in the book. Chapter 6 describes the Ascend/Cascade B-STDX 9000 frame relay switch with ATM uplinking. Both of these chapters show how vendors have approached the support of Quality of Service in ATM and frame relay switches.

The remainder of this book, Chaps. 7 to 11, addresses deployment considerations for the technology.

- Chapter 7 examines the issue of support of IP over an ATM network, a critical requirement in corporate networks.

- Chapter 8 looks at economic considerations for the deployment of ATM.

- Chapter 9 looks at ATM design issues.

- Chapter 10 discusses the challenges faced by carriers in establishing reliable and flexible broadband wide area networks that can be utilized by organizations to extend the enterprise network reach.

- Chapter 11 provides a view of how voice over ATM can be supported to achieve economies of integration and of scale in ATM-based networks. The networks of the mid-to-late 1980s, as well as the drive for multimedia over the Internet, document the advantages of having a single network support as much of an organization's communications requirements as possible.

Acknowledgments

The authors would like to thank Roy D. Rosner, Senior Vice President of Marketing, Telematics, for supplying material for Chap. 5, and Jim Mathison and Jim Martel of Ascend Communications, Inc. for doing the same for Chap. 6.

John Amoss would like to thank his family for its support throughout this endeavor.

Daniel Minoli would like to thank Mr. B. Occhiogrosso, President, DVI Communications, New York, and Ms. J. A. Dressendofer, President, IMEDIA, Morristown, N.J., for their support.

—JOHN AMOSS
—DANIEL MINOLI

IP APPLICATIONS
WITH ATM

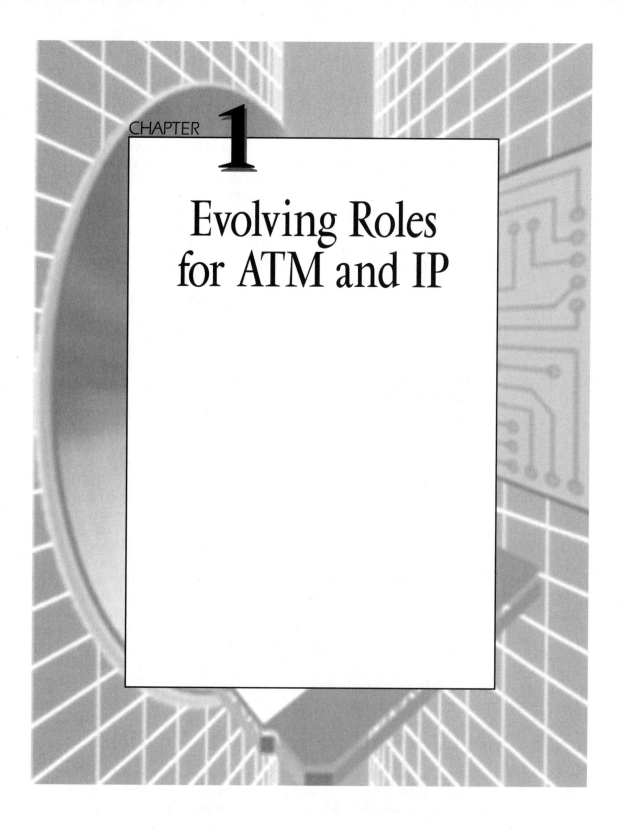

CHAPTER 1

Evolving Roles
for ATM and IP

1.1 Promise of ATM

Substantial work on the specification of asynchronous transfer mode (ATM) was begun in the early 1980s as part of a global effort to define the next infrastructure for public networks. In 1988 this work culminated in an initial series of standards recommendations from the international standards bodies as part of the definition of the Broadband-Integrated Services Digital Network (B-ISDN). Significant debate had resulted in the definition of the basic ATM cell structure so that it would meet the communications needs of various applications—voice, data, etc. A basic transport capability, Synchronous Optical Network/Synchronous Digital Hierarchy (SONET/SDH), was also defined at that time for the carriage of these cells. Work on the specification of this infrastructure continued into the 1990s.

About this same time, the customer premises equipment (CPE) industry was experiencing the need for a next-generation premises networking technology. The technology of the time, based on the IEEE 802 local area network (LAN) standards, operated in the 4, 10, and 16 megabit per second (Mbps) range and was having difficulty supporting the additional bandwidth demands of end users. The technology also lacked inherent Quality of Service features for the support of services such as interactive voice. With ATM, the CPE industry saw an opportunity to provide a premises infrastructure capable of handling voice, video, and data on a single network, with the additional advantage of providing seamless interconnection to the evolving public ATM networks for wide area network (WAN) transport.

The early 1990s were highlighted by a rapid increase in interest in the ATM technology on the part of communications equipment suppliers and service providers. This interest was evidenced by the rapid growth in membership in the ATM Forum, the industry body chartered to hasten the deployment of ATM technology. More than 700 companies are members of the ATM Forum. Currently most of the standards and specifications to enable the deployment of an ATM infrastructure are in place. This book addresses issues related to the realization of the promise of ATM. Such issues involve standardization, switching technology, support of the Internet Protocol (IP) and existing intranets, economic viability, network design, and integrated services support. This book is intended as a guide for corporate and service provider planners.

1.2 Market/Technical Realities

Widespread deployment of ATM faces a number of market and technical realities, some of which should influence its deployment positively and some negatively. Major factors are the following.

1.2.1 Internet Explosion

Internet traffic is projected to grow by more than an order of magnitude annually through the end of the century. Thus, Internet backbone providers are currently in the midst of a major effort to scale up their networks to handle this enormous growth. A number of the 20 to 40 major Internet backbone network providers (including UUNET and MCI) have deployed ATM technology, and a number of others (including AT&T WorldNet) have plans to do so. With this projected traffic growth, the Internet should be a major driver for ATM deployment; the efficient internetworking of IP and ATM technology is a current industry focus.

1.2.2 Technology Evolution

Other high-bandwidth technologies are being introduced in the WAN and LAN environments. In the WAN, wavelength division multiplexing (WDM), a technology that employs multiple wavelengths as carriers on a single fiber, greatly expands the bandwidth capabilities of an installed fiber facility (see Fig. 1.1). The technology will allow more cost-effective carriage of ATM cells and should have a positive impact on the use of ATM in the WAN. In the LAN environment, higher-speed IEEE 802 LANs have been defined, and this is likely to inhibit ATM in this environment. A description of each of these key technologies follows.

Wavelength Division Multiplexing   When they were deployed over the last decade, fiber-optic networks were thought to provide essentially inexhaustible capacity; however, the existing networks are, quite surprisingly, becoming exhausted. SONET (discussed in Chap. 2) has traditionally provided a migration path from Mbps to gigabit per second (Gbps) rates; for example, OC-48 rates, at 2.4 Gbps, are common in current systems.

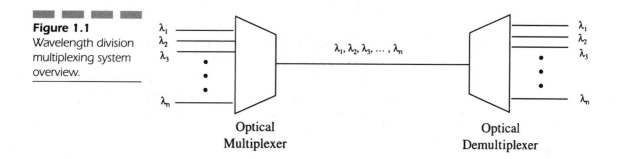

Figure 1.1
Wavelength division multiplexing system overview.

However, with the introduction of OC-192 rates, at 9.6 Gbps, the limits of SONET-based networking have become apparent. At these rates, processing and regenerating signals electrically has posed difficult problems, and cost and technical hurdles associated with OC-192 rates are delaying its market acceptance.

One solution to the bandwidth scarcity problem is WDM, a technique that increases the number of channels on a fiber-optic link through the use of multiple wavelengths on existing light-wave transmission systems and fiber. WDM networks are protocol- and format-transparent, allowing an incremental buildup of electronic networks to optical, as each wavelength acts as a carrier for a channel for any type of service, both SONET-based and non-SONET-based, e.g., analog video.

WDM technology not only increases the number of channels on a given fiber-optic cable, but also can be used as the basis for an optical network. In such a network, the functions of switching, routing, and multiplexing are performed optically rather than electronically. WDM systems can thus reduce the need for new fiber and electronics in many cases and can be the basis for a flexible, highly reliable optical network.

There remain a number of challenges facing the widespread deployment of WDM systems.

■ In contrast to SONET operations and management capabilities, which at this time are mature, WDM capabilities in this area are still in their infancy. Few standards exist to date.

■ WDM systems are still largely deployed to increase bandwidth on point-to-point fiber links. Few optical networking products are currently available.

■ Costs for WDM systems are still high.

In general, WDM should have a positive impact on the deployment of ATM, since it offers more cost-effective cell transport. However, a major advantage of ATM is its bandwidth management capabilities, which, through statistical multiplexing, minimize bandwidth use. WDM will tend to diminish this advantage, since it offers significantly less costly bandwidth.

LAN Evolution (Gbps Ethernet)   The IEEE 802.3z standards activity is extending Ethernet LAN speeds to the Gbps range. Initial efforts drew heavily on the use of available, proven technologies and methods to minimize time to market for Gigabit Ethernet products. For example, integrated circuits and optical components specified in the Fibre Channel

TABLE 1.1

Gigabit Ethernet and ATM in the LAN Environment

	Pros	Cons
Gigabit Ethernet	Relatively inexpensive	Distance limitations over copper and shared nets
	Easy migration path	No Quality of Service
	Same management as standard Ethernet	Standard not available until 1998
ATM	Multivendor products	Requires retraining of staff
	Supports Quality of Service	Relatively expensive
	Reliability	
	Bandwidth management	

Standards, specified by the ANSI X3T11 group, were employed with some modifications.

The IEEE Higher-Speed Study Group has identified three specific distance requirements for Gigabit Ethernet and associated media for these distances.

1. A multimode fiber-optic link with a maximum length of 500 m

2. A single-mode fiber-optic link with a maximum length of 2 km

3. A copper-based Category 5 unshielded twisted pair link with a maximum length of at least 25 m

Table 1.1 contrasts the advantages of Gigabit Ethernet and ATM in the LAN environment.

Currently, with the success of competing technologies such as Fast Ethernet and Gigabit Ethernet, the goal of bringing ATM to the desktop is achieving very limited success. The amount of native ATM encountered on customer premises is modest at best, with the majority of data traffic being offered to carriers in the form of frame relay or private line.

1.2.3 Difficulties with Implementation of Multiservice Networks

While conceptually appealing, the implementation of a single network infrastructure that efficiently carries traffic with vastly differing requirements is an extreme technical challenge. A large number of fairly complex standards need to be defined and implemented so as to allow the interoperabilty of supplier equipment that supports this infrastructure. As mentioned above, this process to date has spanned a period approaching two decades.

1.2.4   Lack of a "Killer Application"

The applications that were expected to drive the technology initially, such as the delivery of video to the home using switched ATM capabilities, have not yet proven cost-effective in the market. Rather, the use of ATM as a backbone for frame relay and IP traffic has led to a moderate demand but has not produced the "killer application" needed to bring about significant price reductions in ATM components. In addition, end users need a clear, demonstrable business justification for deploying new technologies such as ATM, and the lack of a clearly identifiable killer application has hindered this justification.

1.3   Evolving Role for ATM

1.3.1   Drivers

The current state of deployment for ATM technology can best be described as somewhat chaotic. As of this writing, the authors judge the following applications to be the drivers for ATM technology deployment, listed in their order of importance.

- *Frame relay backbones.*   Frame relay traffic continues to grow rapidly, with the U.S. services market projected to grow from approximately $1 billion in 1996 to $10 billion in 2002. As mentioned, this application is dominant today in its use of ATM technology for backbone purposes; continued growth is anticipated.

- *Internet backbones.*   The enormous growth of traffic in this area should present a major opportunity for ATM. Issues related to the transport of IP are addressed in detail in Chap. 7.

- *Infrastructure for new telecommunications entrants.*   For new entrants in a "green field" environment, ATM coupled with SONET/SDH transport capabilities provides a flexible network infrastructure that is ideal for offering a mix of data and potentially voice services.

■ *Backbones for established carriers.* Many established carriers are also deploying ATM in their backbone networks to provide an infrastructure for data and voice services. These large carriers expect to use ATM initially to carry frame relay, private line, and IP traffic. Most carriers anticipate a 2- to 4-year period before the standards for voice are mature and the use of ATM to transport voice services is widespread in public networks. Chapter 11 addresses issues associated with using ATM for the transport of voice traffic.

1.3.2 Evolving ATM Network Architecture

The architecture of an ATM network has evolved toward a tiered structure, depicted in Fig. 1.2. One tier, termed service multiplexing, provides traditional legacy interfaces such as frame relay and Ethernet, conversion to ATM, and a concentration or multiplexing function to deliver traffic to ATM switching tiers. An ATM service multiplexer would typically be located close to, or on, a customer's premises. Service multiplexers are elements that are smaller than edge switches and have a single network port for connection to a public ATM network with any number of subscriber ports. Service multiplexers can be owned and operated by either a carrier or a subscriber. In the latter case, the subscriber would contract for an ATM connection on an ATM service port on an edge or core switch. The

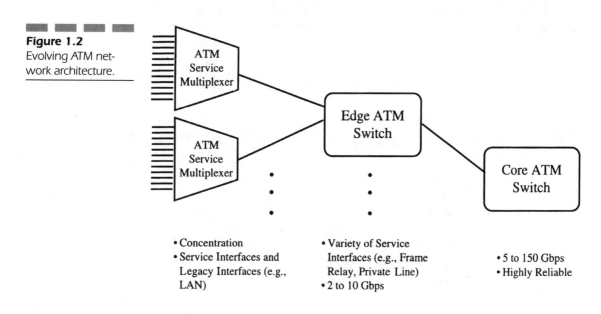

Figure 1.2
Evolving ATM network architecture.

subscriber would use its own service multiplexer to carry a variety of ATM and non-ATM services to and from the public ATM network. Service multiplexers are typically not designed to support full redundancy, as edge and backbone switches are.

Another tier, called edge switching, provides a wide range of traditional non-ATM interfaces along with a first layer of ATM switching capability. Edge switches connect via trunk ports with a backbone switch, and preferably the switch connects to at least two different backbone switches. This allows for the rerouting of connections if one port fails (either at the edge switch or at one of the backbone switches) or if the transmission links fail. Edge switches are viewed as an essential part of a public ATM network and are equipped to provide the same level of system availability as backbone switches. A possible exception to this requirement is subscriber line interfaces. Here, based on the type and purpose of the service provided, individual nonduplicated line interfaces are typically used, unless a premium service with guaranteed high availability is being provided. Throughput of these edge switches is in the 1 to 5 Gbps range, and they may be deployed in a corporate backbone or as a carrier service vehicle.

Many large carriers are moving toward the use of a core or tandem tier in their networks, providing pure ATM-to-ATM switching. These larger switches are currently in the 5 to 25 Gbps range, with major suppliers having either introduced or announced switches on the order of 150 Gbps. A few have announced that their flagship products will be on the order of 600 Gbps in the next few years.

Chapters 2 and 3 present an overview of the ATM communications model and the status of standards associated with the model. The basic architecture of ATM switching elements is addressed in Chap. 4, and detailed examples of supplier products are presented in Chaps. 5 and 6.

1.4 Support of IP over ATM

As a consequence of the popularity of IP one of the key considerations regarding ATM technology is its support of IP. This fact is driven by the need to support the embedded base of applications and enterprise networks (including intranets) and the need to have access to the Internet, including virtual private networks (VPNs).

Classical IP over ATM, developed by the Internet Engineering Task Force (IETF), is one method of moving LAN/intranet traffic over ATM.

The IETF's specifications are documented in the following requests for comments (RFCs):

- RFC1483, *Multiprotocol Encapsulation over ATM Adaptation Layer 5*
- RFC1577, *Classical IP and ARP over ATM*
- RFC1755, *ATM Signaling Support for IP over ATM*
- RFC2022, *Multicast Address Resolution (MARS) Protocol*

These RFCs treat an ATM connection as a virtual "wire" requiring unique means for address resolution and broadcast support. The functionality of address resolution is provided with the help of special-purpose servers and software upgrades on legacy routers. One of the limitations of this approach is that it does not take advantage of ATM Quality of Service features. It also has the drawback of supporting only IP, since the address resolution server is knowledgeable only about IP. This approach does not reduce the use of routers, currently viewed as network bottlenecks. However, the model's simplicity does reduce the amount of broadcast traffic and interactions with various servers; in addition, once the address has been resolved there is the potential that subsequent data transfers may take advantage of this information.

Multicast support is provided via the multicast address resolution server (MARS). The MARS model operates by requiring a multicast server to keep membership lists of multicast clients that have joined a multicast group. A client is assigned to a multicast server by a network administrator at configuration time. The MARS approach uses an address resolution server to map an IP multicast address onto a set of ATM endpoint addresses of the multicast group members.

Beyond this basic support of IP over ATM, the industry is also looking at ways to use the advantages of ATM to simplify IP-level forwarding. Some of the challenges being addressed are (1) support of more than best-effort services on the same router-based network, (2) support of circuitlike services via IP on a router-based network, and (3) support of various service levels, regardless of physical media and network discipline (e.g., LAN switching, traditional IP routing, IP routing over ATM). Users and planners want faster networks with scalability, and better performance and management to accommodate the increased corporate dependence on easy access to information, including data, video, voice graphics, and distributed resources.

A number of these IP themes are addressed in later chapters of this book.

1.5 Economic and Design Considerations

The deployment of ATM will in the final analysis be based on the cost-effectiveness of the technology as compared to other LAN and WAN technologies and its ability to interwork with IP, the dominant networking protocol. Chapter 7 addresses issues related to IP and ATM interworking. Chapter 8 addresses current ATM equipment costs for the service multiplexer, edge, and backbone products. Design considerations and guidelines for engineering and installing an ATM network are presented in Chap. 9, and Chap. 10 presents specific considerations for carrier ATM networks. Chapter 11 discusses issues related to carrying voice traffic over a backbone network and achieving economies of integration.

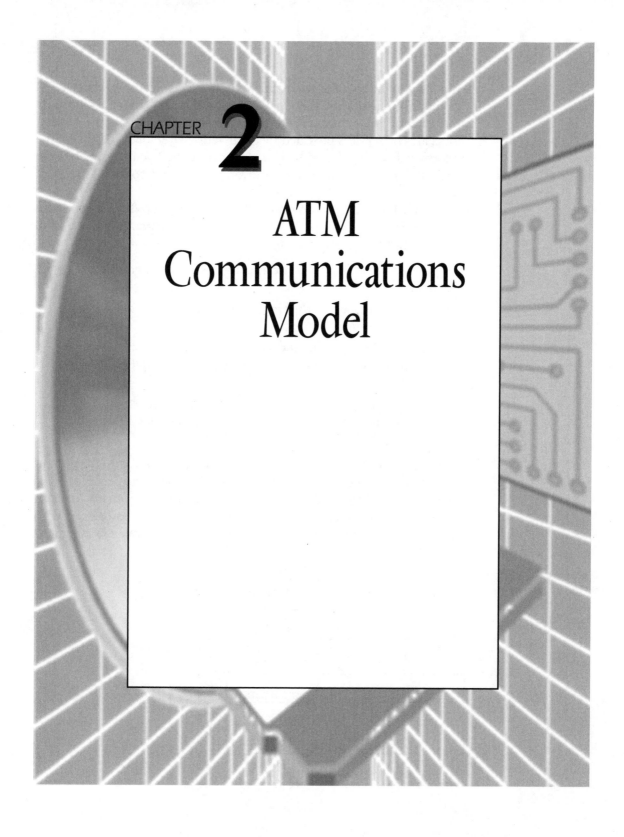

ATM
Communications
Model

This chapter introduces the terminology and concepts used in ATM through a discussion of the layered communications model commonly used in modern communications networks. Inherent in the layered model are the services provided by a particular layer (or set of layers), the performance or Quality of Service required of the layer(s), characteristics of the traffic offered to a layer, and operations and control aspects of the layers.

The ATM layered protocol model, defined in ITU-T Recommendation I.121, is shown in Fig. 2.1. The major layers in the ATM model and their functions are the following.

■ *Physical Layer:* provides basic connectivity between network elements.

■ *ATM Layer:* provides for the transparent transfer of fixed-size (48-octet) data units.

■ *ATM Adaptation Layer:* enhances the services and performance provided by the ATM Layer to meet the needs of higher-layer applications.

■ *Higher layers:* represent various end-user applications, e.g., Transmission Control Protocol/Internet Protocol (TCP/IP) and connection control aspects.

In addition to the above layers, the model contains management planes that interact with these layers. The management planes provide network supervision functions. Layer Management performs management functions relating to resources and parameters associated with layer protocol entities. Plane Management performs management

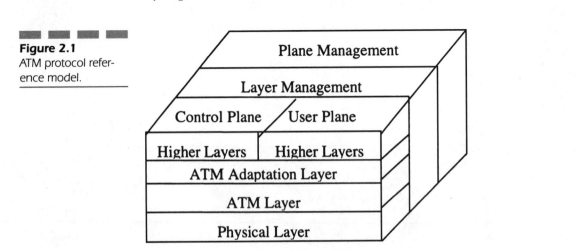

Figure 2.1
ATM protocol reference model.

functions related to a system as a whole and provides coordination among all planes.

This chapter focuses on the general principles of a layered communications model and specifically covers the layers of the ATM protocol reference model: the Physical Layer, the ATM Layer, and the ATM Adaptation Layer.

2.1 Layer Principles

This section introduces the principles and terminology of a layered communications model and focuses on the terminology and sample communications functions common to the ATM protocol architecture. Figure 2.2 indicates the terminology employed for the model, showing a general layer (N) and the two adjacent layers $(N+1)$ and $(N-1)$. Procedures and functions within a layer are performed via protocol entities in the layer. These entities are generally implemented via software in the higher layers of the protocol architecture and in hardware or firmware in the lower layers. Entities communicate with adjacent layers through service access points at interfaces between the layers. As shown in the figure, the procedural rules governing the communication between peer entities, i.e., communicating entities at the same level, are formally termed protocols. Peer entities may reside in the same system or in geographically separate systems. Two peer entities communicating with each other view the entirety of the lower layers as simply providing a connection; thus, as shown, the (N) layer provides an (N) connection to communicating $(N+1)$ entities. For example, in the ATM reference model of Fig. 2.1, the ATM Layer provides an ATM connection to the ATM Adaptation Layer.

The exchange of information between adjacent layers is carried out via what are termed service primitives, essentially subroutine calls with associated primitives. As shown in Fig. 2.3, four basic primitives are used to invoke and provide the results of a service request:

- *Request*—to invoke the service
- *Indication*—to indicate to the peer that a service has been requested
- *Response*—to provide information on how the request was handled (e.g., accepted, rejected)
- *Confirmation*—to provide the final disposition of the request to the originator

Figure 2.2
Layer concepts and terminology.

Query: In text the Ns are italic? Minus signs are hyphens?

Figure 2.3
Sequence diagrams for service primitives.

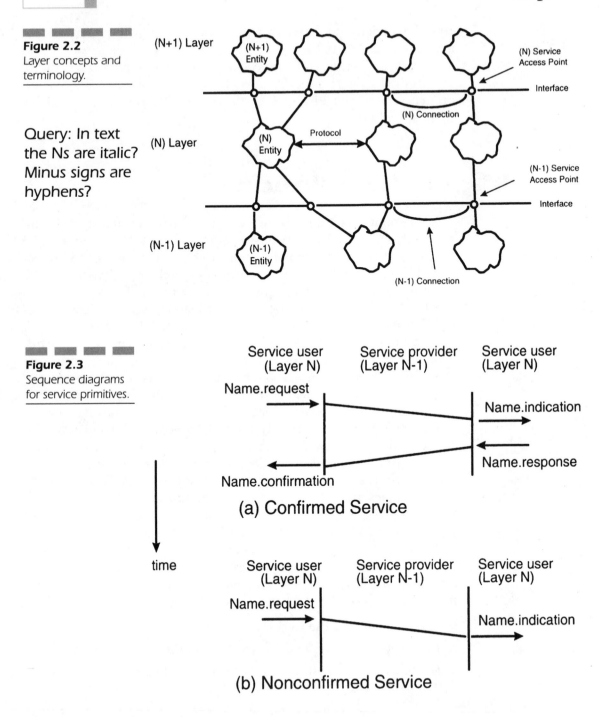

(a) Confirmed Service

(b) Nonconfirmed Service

The term *Name* in the figure refers to the type of action or service invoked, e.g., establishing a connection would involve *connect.request*(), *connect.indication*(), *connect.response*() (e.g., accept, reject, renegotiate,...), and *connect.confirmation*(). Parameters associated with the connection would be included with the request, e.g., *connect.request*(destination address, source address, Quality of Service provided by the connection, priority,...). As another example, sending data is accomplished via the *data.request* primitive, and among the parameters passed would be the actual data to be delivered and a connection identifier. As shown in Fig. 2.3, note that a nonconfirmed service is also defined. Here the results of the service invocation are not returned to the requester; any required subsequent actions, such as those to overcome nondelivery of data, would be accomplished at higher layers.

In addition to the above interactions between entities, it is often helpful to consider the various states of a single-layer entity (e.g., idle, active), along with the allowable transitions between these states and the stimulus for the transitions. A sample state transition diagram is shown in Fig. 2.4. As shown, the states for the entity may be

1. Idle

2. Outgoing connection pending (awaiting the results of a *connect.request*)

3. Incoming connection pending (having received a *connect.indication* and not yet replied)

4. Data transfer ready (a state in which *data.request, data.indication, data.response,* and *data.confirmation* primitives are sent and received)

These state transition diagrams are helpful in the protocol design phase and can be used to identify states that cannot be reached or that, once reached, cannot be exited. Note that these diagrams apply to only a single-layer entity; a complementary diagram will be in effect for the peer entity involved in the communications process.

Finally, we conclude this overview of layered communications principles with a discussion of how a layer entity actually performs the functions for which it is responsible and a description of the implementation of two functions common in the ATM architecture, segmentation/reassembly and multiplexing/demultiplexing.

Figure 2.4
Sample state transi-
tion diagram.

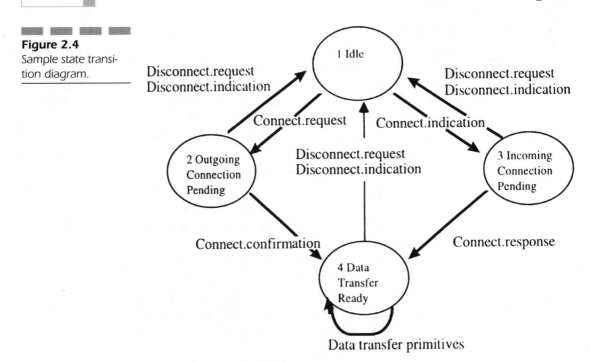

2.1.1 Protocol Control Information

The generic tool used by a layer entity is the addition of protocol control information (PCI), commonly termed header information, to the information received from the higher layer. This PCI contains fields to perform the layer functions. Figure 2.5 indicates the case in which layer (N) receives information from layer $(N+1)$, termed the service data unit (SDU), and attaches its layer (N) PCI to this information. The result is termed the (N) layer protocol data unit (PDU). Note here that the only layer to interpret the layer (N) PCI is the peer end layer in the far end communicating system; lower layers will treat the (N) PCI as part of the information passed from above.

2.1.2 Segmentation/Reassembly

Figure 2.6 indicates the case in which layer (N) receives a data unit from layer $(N+1)$ whose length exceeds the maximum allowable length of data units carried on the connection provided to it by layer $(N-1)$. This is a common occurrence in layered architectures, e.g., an X.25

network is often limited to 256- or 1024-byte packets, Ethernet frames are limited to 1518 bytes, and an ATM connection accepts only 48-byte data units. The function performed by the (N) layer entity is thus the segmentation of the data unit into smaller units that are able to be carried by the lower-layer connection. Note that its peer will need to reassemble the original data unit from these smaller units, the reverse of the process shown in Fig. 2.6. Thus the required PCI generally includes a sequence number to enable the receiving peer to correctly order the received data units for reassembly and an indication as to whether this is the first of the smaller data units, one in the middle, or the last. Providing the indication that this is the last data unit in the sequence allows the receiver to begin the reassembly process (assuming that all data units have been received).

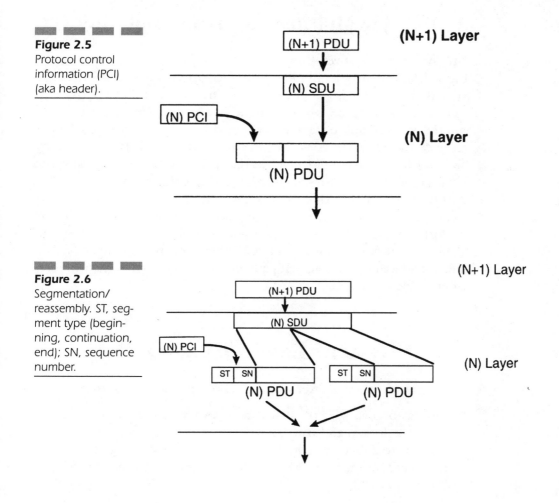

Figure 2.5
Protocol control information (PCI) (aka header).

Figure 2.6
Segmentation/reassembly. ST, segment type (beginning, continuation, end); SN, sequence number.

Figure 2.7
Layer multiplexing/
demultiplexing. MID,
multiplexing ID
[identifies PDUs
belonging to a partic-
ular (N) connection].

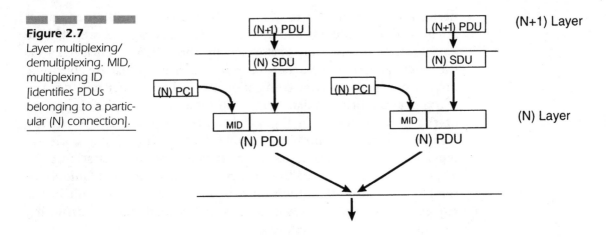

2.1.3 Layer Multiplexing/Demultiplexing

Another common layer function is the use of a single connection to carry data units from a number of higher-layer connections. This situation is shown in Fig. 2.7, where data units from two layer (N) connections are carried over a single layer $(N-1)$ connection. Such multiplexing/demultiplexing is performed in a number of layered architectures, including the X.25 layer, the TCP layer in the TCP/IP protocol architecture, and the ATM Layer in the ATM model. To accomplish this function, the layer entity includes a multiplexing identifier (MID) as part of its PCI. This identifier is assigned by the originating layer entity to designate a particular connection and is used by the layer's peer to perform the demultiplexing function. Examples of MIDs include the 12-bit virtual circuit number of X.25, the 16-bit port number of TCP, and the 8-bit and 16-bit virtual path identifier and virtual channel identifier of the ATM layer.

2.2 Physical Layer (e.g., SONET)

In general, the physical layer performs two types of functions.

1. Those associated with the structure or format of the information to be transmitted and with specific functions performed by the physical layer, such as multiplexing. These functions are grouped into what is termed the *Transmission Convergence* (TC) Sublayer, as shown in Fig. 2.8.

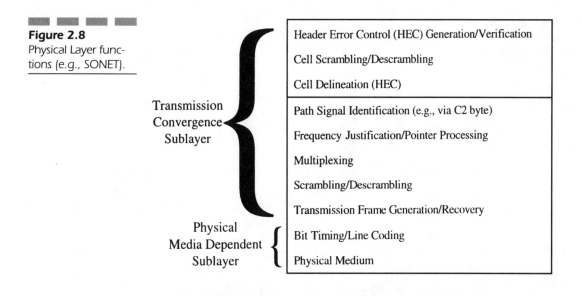

Figure 2.8
Physical Layer functions (e.g., SONET).

Transmission Convergence Sublayer
{
Header Error Control (HEC) Generation/Verification

Cell Scrambling/Descrambling

Cell Delineation (HEC)

Path Signal Identification (e.g., via C2 byte)

Frequency Justification/Pointer Processing

Multiplexing

Scrambling/Descrambling

Transmission Frame Generation/Recovery

Physical Media Dependent Sublayer
{
Bit Timing/Line Coding

Physical Medium

2. Those associated with the transmission of signals over a particular medium (fiber, coax, twisted pair). These functions are associated with a *Physical Media Dependent* (PMD) Sublayer. An example of such functions is the specification of the line coding for the particular medium, e.g., generally non-return-to-zero (NRZ) for fiber, bipolar with three zero substitution (B3ZS) for coax, and NRZ for twisted pair.

Such a structure has the obvious advantage of reuse of functionality; transmission convergence functions can be defined independently of the type of medium employed, and subsequently used over a variety of physical media.

For an ATM network, the TC Sublayer generally uses the structure of the Synchronous Optical Network (SONET) [ANSI T1.105] [GR-253] or the similarly defined Synchronous Digital Hierarchy (SDH) [G.707] [G.708] [G.709]. As will be discussed in Chap. 3, the transport of ATM cells is being defined for a wide range of physical media (e.g., optical fiber, twisted copper pair wire) and transmission rates (e.g., 25 Mbps, 51.840 Mbps, 155.520 Mbps). In this section, we discuss the SONET rates and formats in some detail and briefly address a primary medium used in the local area, Category 5 twisted pair wire.

2.2.1 SONET Overview

While SONET provides a specification for the Physical Layer of the ATM protocol model, its functions are also defined according to the layering

Figure 2.9
SONET layers.

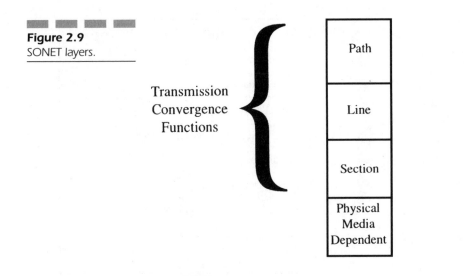

Transmission Convergence Functions

Path

Line

Section

Physical Media Dependent

principles outlined in Sec. 2.1. Its four functional layers, shown in Fig. 2.9, include the following.

- The PMD Sublayer, discussed above
- The Section Layer, concerned with the transport of the SONET frame across the physical medium
- The Line Layer, concerned with functions such as protection switching (to protect against line failures) and multiplexing
- The Path Layer, concerned with the transport of various types of payload, such as DS1 signals, DS3 signals, and ATM cells

With the exception of the PMD Layer, each of these layers has associated protocol control information (section overhead, line overhead, and path overhead) used to perform the layer functions. In SONET terminology, the section and line overheads are collectively referred to as the transport overhead.

2.2.2 SONET Rates and Formats

A SONET-based physical layer interface utilizes the frame structure shown in Fig. 2.10. This frame, referred to as the Synchronous Transport Signal-1 (STS-1), is generally depicted as a collection of bytes arranged in rows and columns, with transmission proceeding from left to right, top to bottom. As shown in the figure, the nominal bit rate is 51.840 Mbps, with $90 \times 9 = 810$ bytes clocked out every 125 μs. The Line Layer payload

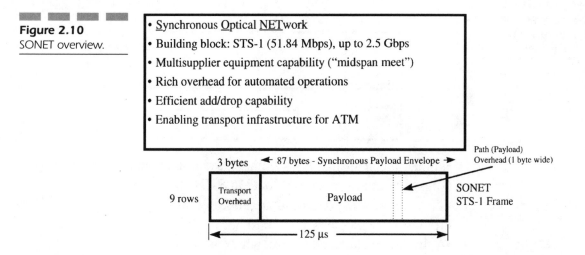

Figure 2.10
SONET overview.

- <u>S</u>ynchronous <u>O</u>ptical <u>NET</u>work
- Building block: STS-1 (51.84 Mbps), up to 2.5 Gbps
- Multisupplier equipment capability ("midspan meet")
- Rich overhead for automated operations
- Efficient add/drop capability
- Enabling transport infrastructure for ATM

capacity, termed the STS-1 Synchronous Payload Envelope (SPE), is also shown in the figure. The STS-1 frame structure also contains fixed undefined stuff bytes (not shown in the figure and not usable for user information) in columns 30 and 59. Thus, the bit rate available for user information (e.g., ATM cells), excluding information transported in the SONET overhead bytes and the fixed stuff bytes, is 48.384 Mbps.

2.2.3   SONET Pointer Mechanism

To account for the possibility of a payload rate being slightly out of synchronization with the line rate, SONET allows the line payload to "float" within the SONET payload envelope. This is accomplished through the use of a pointer mechanism, implemented as part of the line overhead, to locate the start of the payload structure (see Fig. 2.11). As shown, the line overhead also provides a byte to accommodate frequency adjustment when the frame rate of the SPE is greater than that of the STS-1. Procedures are also defined to adjust for the frame rate of the STS-1 being less than that of the SPE.

2.2.4   SONET Layer Functions and Protocol Control Information

Section and line overhead are indicated in Figs. 2.12, 2.13, and 2.14. The primary function of the Section Layer is to provide overall STS-1 framing

Figure 2.11
Pointer mechanism.

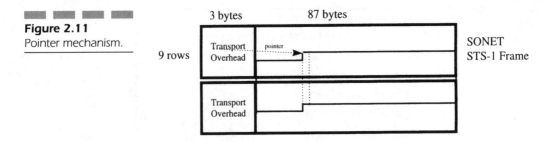

•Pointer mechanism allows frames carried in payload to be (slightly) out of sync; can "float" in STS-1 frame
•Two bytes of line overhead are allocated to the pointer (10 bits carry pointer value - 0 to 782)
•A third byte allocated for frequency adjustment

(accomplished via the A1 and A2 bytes). The Section Layer also uses a common technique, termed *scrambling/descrambling,* to assist in deriving clock timing from the input signal. Essentially, the technique randomizes the signal bits at the transmitter in a manner that is reversible at the receiver, so that the likelihood of a string of ones or zeros (and thus potential loss of timing) is reduced. Scrambling/descrambling operates at the line rate, and the framing bytes and the C1 byte are not scrambled (see Fig. 2.12). Note that scrambling/descrambling can also be employed at other layers; for example, at the ATM Layer, this technique is employed on the ATM cell payload, since physical layers other than SONET may be used. Error detection at the Section Layer is provided via a parity check (B1 byte).

As discussed, the Line Layer provides alignment of the SPE payload (via the H1 and H2 bytes) and frequency adjustment (via the H3 byte). Error detection is also provided at this layer (via the B2 parity byte). The Line Layer provides a protection switching function for a number of physical configurations (point-to-point and ring), and this function is accomplished via a protocol using the K1 and K2 bytes. Finally, a separate channel for the exchange of operations, administration, and maintenance (OAM) information is provided via the D4 to D12 bytes. This 576-kbps channel provides physical connectivity for the exchange of OAM information among SONET network elements.

The path overhead (see Figs. 2.15 and 2.16) provides additional error detection via a parity byte (B3). SONET is defined as a general-purpose transport mechanism and as such is designed to carry various payloads, with the Path Layer responsible for mapping such payloads to the SPE. The particular

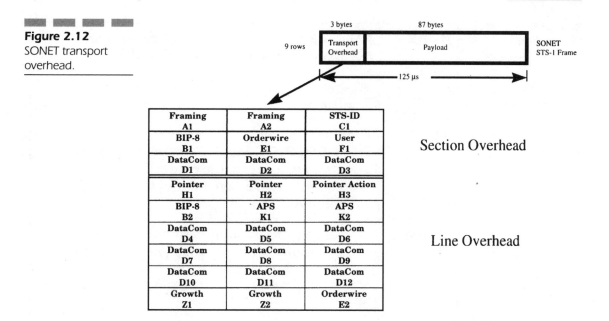

Figure 2.12
SONET transport
overhead.

Section Overhead

Line Overhead

Figure 2.13
Section overhead
bytes.

A1, A2	Framing bytes = F6,28 hex; used to synchronize the STS-1 SONET frame
C1	Identifies the STS-1 number (1 to N) for each STS-1 within an STS-N multiplex
B1	Bit-interleaved parity byte providing even parity over previous STS-N frame
E1	Voice channel to be used between section terminating equipment
F1	64-kbps channel for user purposes
D1-D3	192-kbps data communications channel for alarms, maintenance, control and administration between section terminating SONET equipment

Figure 2.14
Line overhead bytes.

H1-H3	**Pointer bytes used in frame alignment and frequency adjustment of payload data**
B2	**Bit-interleaved parity for line level error monitoring**
K1, K2	**For signaling between line level automatic protection switching equipment; uses a bit-oriented protocol that provides for error protection and management of the SONET optical line**
D4-D12	**576-kbps data communications channel for alarms, maintenance, control, monitoring and administration at the line level**
Z1, Z2	**Reserved for future use**
E2	**Line level 64-kbps voice channel**

mapping is indicated by the C2 byte. For example, a value of 00010011 indicates the carriage of ATM cells; 00000010 indicates the carriage of virtual tributaries (discussed below), which could in turn carry, e.g., DS1s.

2.2.5   SONET Payloads

Lower-Rate Signals

Virtual Tributaries   Signals that have a lower rate than the STS-1 signal, such as DS1 signals at 1.544 Mbps, are carried in the SPE in what are termed virtual tributaries (VTs). A number of VT structures are defined: VT1.5 to transport DS1 signals; VT2 to transport 2.048-Mbps signals (a common signal rate in Europe); and VT3 and VT6 to carry 3.153-Mbps and 6.312-Mbps signals, both of which are seldom used. The VT structures include another set of overhead bytes providing functions similar to those associated with the section, line, and path, including frequency adjustment for signals carried in the VT and error detection via a parity check. Figure 2.17 shows the sizes of these VTs, including both the transported signal and the VT overhead.

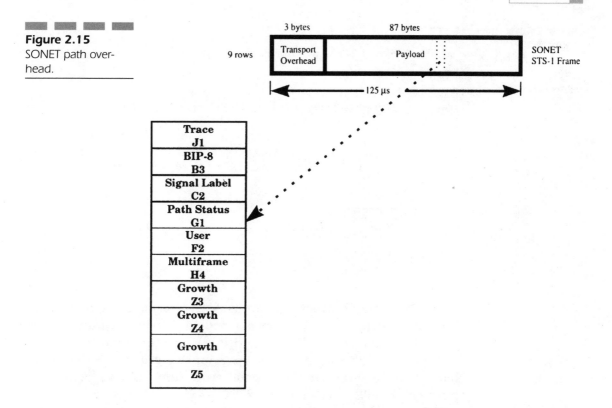

Figure 2.15
SONET path overhead.

Figure 2.16
Path overhead bytes.

J1	64-kbps channel to repetitively send a 64-octet fixed-length string so a receiving terminal can continuously verify the integrity of a path; contents are user programmable
B3	Bit-interleaved parity at the path level, calculated over all bits of the previous SPE
C2	Indicates the specific STS payload mapping, e.g., 00010011 indicates ATM cells in payload; 00000010 indicates VT structure
G1	Status byte sent from path terminating equipment to convey status of terminating equipment and path error performance
F2	64-kbps channel for path user
H4	Multiframe indicator to assist in mapping specific types of payloads into the SPE
Z3-Z5	Reserved for future use

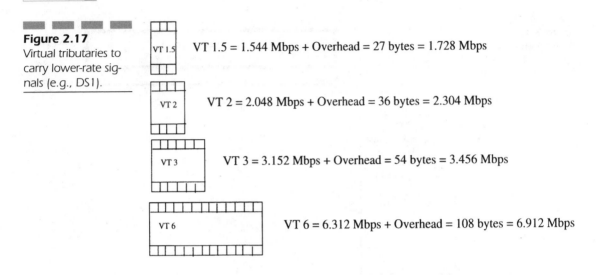

Figure 2.17
Virtual tributaries to carry lower-rate signals (e.g., DS1).

VT 1.5 = 1.544 Mbps + Overhead = 27 bytes = 1.728 Mbps

VT 2 = 2.048 Mbps + Overhead = 36 bytes = 2.304 Mbps

VT 3 = 3.152 Mbps + Overhead = 54 bytes = 3.456 Mbps

VT 6 = 6.312 Mbps + Overhead = 108 bytes = 6.912 Mbps

Packing these VT structures into the SPE is accomplished in two steps. First, combinations of VTs are defined to form VT groups of size $12 \times 9 = 108$ bytes (see Fig. 2.18). Note that a VT group cannot have different types of VTs within the group. Seven VT groups are then byte-interleaved (i.e., bytes are taken sequentially from each VT group) and placed in the SPE as shown in Fig. 2.19.

VT groups carrying different lower-rate signals may be carried in the SPE. As an example, a single SPE could carry the following:

$$4 \times 7 = 28 \text{ VT1.5s} \quad or \quad 3 \times 7 = 21 \text{ VT2s} \quad or \quad 2 \times 7 = 14 \text{ VT3s}$$

$$or \quad 1 \times 7 = 7 \text{ VT6s}$$

Another possibility would be the following:

$$4 \times 4 = 16 \text{ VT1.5s} \quad and \quad 3 \times 1 = 3 \text{ VT3s} \quad and \quad 1 \times 2 = 2 \text{ VT6s}$$

ATM Cells In addition to the VT structures above, and most important for our discussion, the SONET SPE is also used to carry fixed 53-byte ATM cells (discussed in Sec. 2.3). ATM cells are directly mapped into the STS-1 SPE payload capacity by aligning the byte structure of every cell with the byte structure of the SONET frame. The entire payload capacity is filled with cells, row by row. Because the payload capacity is not an integer multiple of the cell length, a cell may cross the SPE boundary.

For SONET interfaces carrying ATM cells, additional functions, cell delineation and cell rate decoupling, need to be performed.

Cell Delineation   It is necessary for the receiver to determine the location of the 53-byte cells within the SPE. This function is performed by determining when a checksum byte, the Header Error Check (HEC) byte, associated with the ATM cell header is valid. This cell delineation process has three states of operation: Hunt state, Presync state, and Sync state.

In an initial *Hunt* state, the delineation process is performed by checking byte by byte for the correct HEC for the assumed header field. Once such an agreement is found, it is assumed that one header has been

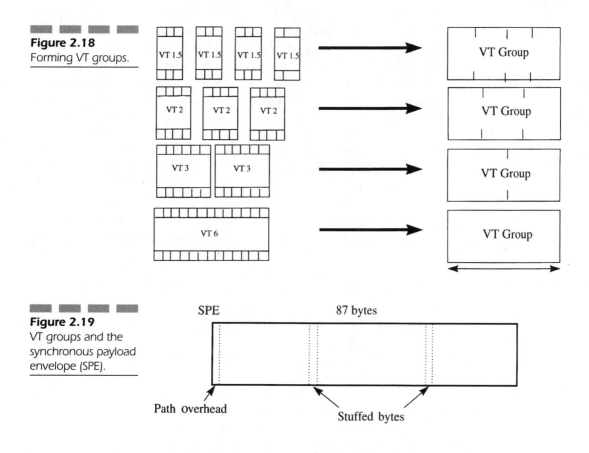

Figure 2.18
Forming VT groups.

Figure 2.19
VT groups and the synchronous payload envelope (SPE).

- Nine bytes are allocated for path overhead
- Two additional columns (18 bytes) are stuffed (undefined) bytes
- Remaining 84 columns carry 7 VT groups, byte-interleaved within the SPE
- Different type groups can be carried within the SPE, but a VT group cannot have mixtures of VTs within the group

found, and the process enters the *Presync* state. In the Presync state, correct HECs are sought on a cell-by-cell basis. If one incorrect HEC is detected, the process returns to the Hunt state. If a preconfigured number of consecutive correct HECs are obtained while in the Presync state, the process enters the *Sync* state. In the Sync state, cell boundaries are assumed to be correct, and correct HECs are verified on a cell-by-cell basis. The process returns to the Hunt state when a preconfigured number of consecutive incorrect HECs are obtained. The standards suggest default values for the preconfigured parameters discussed above.

Cell Rate Decoupling   The Physical Layer expects cells to arrive from the ATM Layer at a rate exactly equal to the payload capacity of the SONET frame (e.g., 48.384 Mbps/424 bits per cell, or just over 114 kcells/s for STS-1). Two techniques are available to ensure that sufficient cells are available. (1) The ATM layer can insert dummy cells, termed "unassigned cells," when cells are not available to fill the payload capacity. Unassigned cells are identified by a reserved connection identifier field (VPI=0 and VCI=0), discussed in Sec. 2.3.1. (2) Alternatively, the Physical Layer may insert such cells, termed "idle cells," when sufficient cells are not available from the ATM Layer. Idle cells are identified by a standardized pattern for the cell header.

Higher-Rate Signals   SONET also defines methods for the transport of higher-rate signals. Two methods are employed. In the first method, a higher-rate format is obtained by byte-interleaving multiple STS-1 frames and clocking them out in 125 μs. This procedure is depicted in Fig. 2.20 for three STS-1 frames, resulting in what is termed an STS-3 signal format. The resulting signal rate is 3×51.840 Mbps=155.520 Mbps. Note here that the signal consists of separate and independent STS-1s that can be recovered by the receiver. For example, three pointers, one for each of the STS-1s, are in effect for the STS-3; there are also separate path overhead bytes for each of the STS-1s.

The second method, termed concatenation, is often used to carry ATM cells at higher rates. This format is also used to carry signals whose rate is higher than the STS-1 rate, such as the Fiber Distributed Data Interface (FDDI) signal at effectively 125 Mbps. As an example, a common concatenated format, the SONET STS-3c [ANSI T1.624], is shown in Fig. 2.21. This STS-3c frame structure makes the full payload available to communicating entities on an end-to-end basis. This format requires only one active pointer and requires only a single path overhead, rather than the three for the STS-3 signal. Thus the payload available for carrying ATM cells is

$$[270 - 9 \text{ (for transport overhead)} - 1 \text{ (for path overhead)}] \times 9 \text{ bytes}/125 \text{ μs}$$

$$= [260 \times 9 \text{ bytes}/125 \text{ μs}] = 149.760 \text{ Mbps}$$

The most common multiples of the STS-1 rate currently supported by the industry are 3, 12, and 48, yielding STS-3/3c, STS-12/12c, and STS-48/48c.

One final bit of terminology: The STS-1 is the basic SONET format and can be carried on electrical or optical physical media. The optical signal is referred to as Optical Carrier—level 1 (OC-1). Similarly, the

Figure 2.20

Carrying higher-rate signals—STS-3 (155 Mbps).

• Frame structure is a sequence of 3 x 810 bytes formed by byte-interleaving 3 STS-1 frames

•Note that three separate STS-1s are carried in the signal (3 pointers are active)

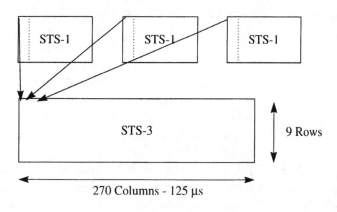

Figure 2.21

Carrying higher-rate signals—STS-3c (155 Mbps).

• Full payload available to end user; not viewed as separate STS-1s.
• Only one pointer active, others simply indicate concatenated signal; only single path overhead required (not three as with STS-3).

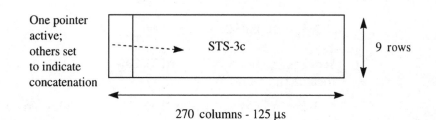

terminology for higher-rate optical signals is OC-3/3c, OC-12/12c, and OC-48/48c.

2.2.6 LAN Physical Media

One medium defined for use by ATM in the local area employs the 155-Mbps Physical Medium Dependent (PMD) Interface using unshielded twisted pair (UTP) Category 5 copper wiring. Category 5 wiring has its spectral characteristics specified up to 100 MHz, whereas Category 3 is defined only to 16 MHz. Category 5 media are intended for use in high-performance networks and provide point-to-point communication between ATM devices and ATM network equipment. Specifications for this medium are provided in the Commercial Building Telecommunications Cabling Standard [ANSI/EIA/TIA-568A-95]. The 568A standard describes performance criteria for cables from both a physical and an electrical standpoint. The performance exhibited by cables varies greatly depending on the frequency or transmission speed of the signal that is carried, and at the multimegabit signaling rates that are required to support ATM, a digital signal on a twisted pair cable deteriorates in signal quality over a fairly short distance. The 568A standard describes the requirements for unshielded twisted pair transmission performance for cables that have a level of performance appropriate for applications in the 100-Mbps range. A 90-m limit is described in the standard as the maximum distance between a telecommunications outlet and the wiring closet. The standard describes the various cabling media that can be used, including 100-Ω UTP cable in a four-pair configuration, 150-Ω shielded twisted pair cable in a two-pair configuration, 50-Ω coaxial cable, and two-fiber 62.5/125-μm optical fiber cable.

2.3 ATM Layer

2.3.1 Services

Services provided by the ATM Layer are rather straightforward. They include the following:

- Transfer of fixed-size (48-byte) data units between communicating upper-layer entities. This transfer occurs on a preestablished ATM connection. The connection process is initiated manually via a paper "service order" process in the case of a permanent virtual connection (PVC) or via signaling in the case of a switched virtual connection (SVC).

■ Multiplexing of a number of ATM connections onto a single physical connection.

■ Ability to differentiate between two types of cells (used, for example, by upper-layer entities to indicate whether or not a cell is the last of a sequence of cells associated with a higher-layer frame).

■ Ability to mark certain data units as more important than others (i.e., two levels of priority).

■ Ability to indicate if congestion was experienced as a data unit passed through network nodes.

The above services are accomplished by use of various header fields added by the ATM Layer entity to the data unit received from the higher layer (AAL). The structure of the ATM protocol data unit is shown in Fig. 2.22.

Examples of the use of header fields in providing the above services are the following.

■ Multiplexing of a number of ATM connections onto a single physical link is accomplished via the VPI/VCI addressing fields shown in the figure. The hierarchical nature of these fields actually allows for two levels of multiplexing: The VCI field allows multiple ATM connections to be multiplexed onto a single virtual

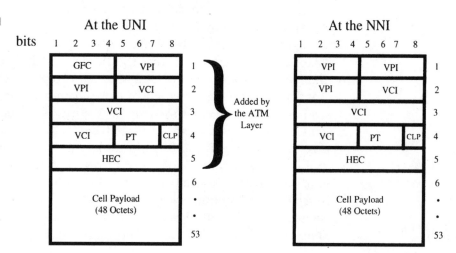

Figure 2.22
ATM cell structure. GFC, generic flow control; VCI, virtual channel identifier; CLP, cell loss priority; UNI, user network interface; PT, payload type; VPI, virtual path identifier; HEC, header error control; NNI, network-to-network interface.

path, and the VPI field allows multiple virtual paths to be multiplexed onto a single Physical Layer connection.

■ Marking certain data units as more important than others is accomplished via the cell loss priority (CLP) field (0=high priority, 1=low priority).

■ ATM switches participating in the ATM connection can indicate congestion (e.g., when switch buffer occupancy has reached a certain critical level) via a congestion indication bit, termed the explicit forward congestion indicator (EFCI), in the cell header (included in the PT field). This information is eventually passed to a management entity in the ATM Layer and used as part of the traffic management process.

■ One bit in the PT field is used to indicate data unit type and is used, for example, to designate a cell as the last one carrying information for a particular AAL PDU.

■ One header byte, the header error control byte, is used for error control of the header fields. As described above, this field is also used to delineate the ATM cells when they are carried in a SONET payload.

It is important to note that the ATM Layer provides the above services to higher layers in the user plane via "user" PDUs, and also to the management layers via operations, administration, and maintenance (OAM) PDUs. A code point in the PT field indicates whether the cell is carrying user information or management information, and in the case of management information, whether the OAM function is on an end-to-end basis or a local (segment) basis. The complete set of code points for the PT field is shown in Table 2.1.

Figure 2.23 depicts the format of the OAM cells originating in the management layers. Note that the headers of these cells (first 5 bytes) are identical to the header of a user cell except for the PTI indication. Particular OAM cells (termed F5 flows) have been defined to deal with end-to-end and switch-to-switch OAM aspects of the ATM virtual circuit connections and are indicated by PT code points. As shown in Fig. 2.23, code point 100 indicates switch-to-switch flows and code point 101 indicates end-to-end cell flows. For virtual path connections, end-to-end and switch-to-switch OAM aspects are indicated via reserved virtual circuits within the virtual path, with VCI=3 used for switch-to-switch cell flows and VCI=4 used for end-to-end cell flows. The particular OAM cell type (fault management, performance

TABLE 2.1	**PTI Coding**	**Interpretation**
Payload Type Indicator (PTI) Code Points	000	User data cell, congestion not experienced, SDU_type=0
	001	User data cell, congestion not experienced, SDU_type=1
	010	User data cell, congestion experienced, SDU_type=0
	011	User data cell, congestion experienced, SDU_type=1
	100	Segment OAM F5 flow-related cell
	101	End-to-end OAM F5 flow-related cell
	110	Traffic control and resource management
	111	For future use

Bit 1: user, OAM, or other; bit 2 (user): congestion; bit 3 (user): type.

Figure 2.23
OAM cell format.

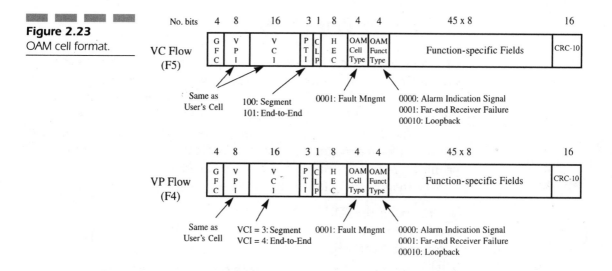

management,...) is indicated in the cell payload via an OAM *cell type* field, with the remainder of the payload available for additional information related to the particular management function. Examples are shown in the figure. Figure 2.24 indicates the details of two maintenance-related OAM cells, AIS/FERF and Loopback.

Finally, cells destined for higher-layer applications in the control plane in Fig. 2.1 are carried on particular virtual channel connections. For example, for point-to-point signaling, this connection identifier has the value VCI=5.

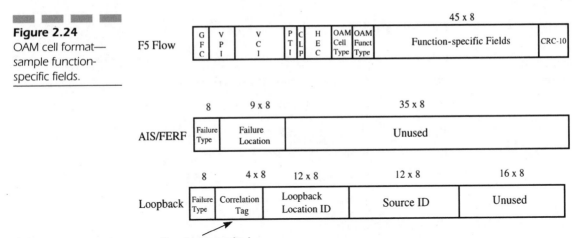

Figure 2.24
OAM cell format—
sample function-
specific fields.

2.3.2 ATM Primitives

As mentioned in Sec. 2.1, layers exchange information via the invocation of subroutines, termed primitives, and parameters associated with the primitives. Table 2.2 lists the primitives used to exchange information between the ATM Layer and the higher adjacent layer (AAL). The primitives shown in the table are those used to exchange payload data carried in the ATM cell and other parameters associated with the cell (e.g., loss priority). Other primitives, not addressed here, are associated with call setup and teardown.

Note the following for the ATM primitives and parameters:

■ One use of the SDU type parameter is to allow the higher layer to signal whether or not this SDU is the last in the message and reassembly can begin.

■ The sending higher layer requests a certain cell loss priority. Actions taken within the ATM network, for example as part of traffic management and congestion control (discussed in Sec. 4.6), may change this priority. The received loss priority can thus be different from the loss priority requested.

■ The congestion experienced parameter is an indication of congestion experienced while in transit through the ATM network and thus is not passed by the sender. If this parameter indicates to the receiver that congestion has been experienced in the network, the receiver is expected to take action to limit its traffic.

TABLE 2.2

ATM Primitives and
Parameters

Primitives	Parameters	Meaning	Values
ATM-DATA.request	ATM SDU	48 bytes for transport over ATM connection	Any 48-byte pattern
	SDU type	Cell type indicator	0 or 1
	Submitted loss priority	Requested cell loss priority	Hi or Lo
ATM-DATA.indication	SDU	48 bytes received over ATM connection	Any 48-byte pattern
	SDU type	Cell type indicator	0 or 1
	Received loss priority	Received cell loss priority	Hi or Lo
	Congestion experienced	Congestion indication	True or False

2.3.3   Quality of Service

While, as discussed above, the basic cell transport services provided by the ATM Layer are straightforward, the complexity of this layer arises from the need to simultaneously support traffic with a variety of performance needs or differing qualities of service. Traffic with both stringent and minimal performance requirements must be managed so as to not interfere with the ATM network's ability to meet the needs or Quality of Service requirements of all traffic classes.

Quality of Service (QoS) refers to a collection of *performance parameters* whose values have to do with the speed or accuracy/reliability characteristics of the ATM connection. For example, for ATM switched virtual connections, requested QoS parameter values are passed along as part of the connection setup message. The ATM Layer will either provide a requested QoS or indicate to the requesting layer that a lesser QoS can be provided.

Typical QoS performance parameters describe transit delay (maximum and average values may be specified) and error rates (e.g., the desired maximum rate of lost data units). ATM QoS is formally defined in terms of *cell transport outcomes* observed at network ingress and egress measurement points [I.356]. Possible outcomes include:

■ Successful transport—cell received within a specified time and with no errors

■ "Tagged cell"—successful transfer, but the cell priority was changed

■ Errored cell—information field changed or header invalid

■ Lost cell—cell not delivered within the appropriate time frame

■ Misinserted cell—a cell delivered with no corresponding input cell

■ Severely errored cell block—a certain percentage of lost, errored, or misinserted cells observed in a block of transmitted cells

Based on the above outcomes, one can define *ATM performance parameters* for cells transported between the measurement points. These parameters are variables that can be measured and for which objective values can be identified and used as part of a "traffic contract" between the ATM network and the user. For example, a cell transfer delay (CTD) could be defined based on the time from the occurrence of an event at the originating network interface (e.g., the beginning of cell transmission) to the corresponding event at the terminating network interface.

Currently, the ATM Forum specifies the following ATM Layer QoS performance parameters to be negotiable between the network and the user.

■ *Peak-to-peak cell transfer delay variation (ppCDV).* Variability of the cell transfer delay. The term *peak-to-peak* refers to the difference between the best- and worst-case CTD, where the best case is equal to a fixed minimum possible delay, and the worst case is equal to a value likely to be exceeded with some small, predefined probability.

■ *Maximum cell transfer delay (maxCTD).* Value of CTD that has only a small, predefined probability of being exceeded.

■ *Cell loss ratio (CLR).* Ratio of total lost cells to total transmitted cells. This ratio can be defined for the subsets of high-priority cells, low-priority cells, and combined cells.

Values for the following parameters are generally beyond the control of the ATM Layer and are therefore not negotiable.

■ *Cell error ratio (CER)*—ratio of errored cells to total cells

■ *Cell misinsertion rate (CMR)*—rate of misinserted cells

■ *Severely errored cell block ratio (SECBR)*—ratio of severely errored cell blocks to total cell blocks

Note that the use of particular QoS performance parameters and the objectives set for the parameters will ultimately depend on the performance requirements of the higher-layer application. For example, an application that is relatively tolerant of cell loss may not include CLR as part of its traffic contract (or may specify a rather loose objective), whereas one that is loss-intolerant would include a stringent objective for CLR. As an example, Bellcore's Broadband Switching System (BSS) Generic Requirements specify a CLR of less than 10^{-10} *for a single switch* in order to meet the needs of the most stringent applications [GR-1110]. CLR requirements for the less stringent applications are specified as less than 10^{-7} *for a single switch*. Of course an ATM connection can involve a number of BSSs and other components that can contribute additional loss on the connection.

Note that the parameters apply to individual ATM connections but need to reflect the impact of various ATM Layer functions. For example, multiplexing multiple ATM connections onto a single physical connection will introduce additional variability in the delay associated with an individual ATM connection. This general situation is shown in Fig. 2.25, extracted from [I.371]. Note here that the information flow at the physical level consists of a mixture of cells containing information from higher layers being carried over ATM connections (the shaded PDUs), and information generated *within* various layers, e.g., F4 and F5 OAM cells generated within the ATM Layer. OAM information from the physical level, e.g., SONET overhead, is also shown in the figure.

2.3.4   Traffic Parameters

A higher layer may also indicate, e.g., via traffic parameter values in a connection setup message, the characteristics of the traffic to be offered over the ATM connection. Traffic parameters indicate a particular aspect of the traffic characteristics and may be qualitative or quantitative. For example, the requesting layer could indicate the maximum, minimum, and average throughput values that it expects to produce; the burstiness of the traffic; and the length of bursts at the maximum rate. The values of these parameters (along with the requested QoS) may be used in determining whether to accept or reject the call and to subsequently monitor whether the traffic on the connection adheres to a "contract" based largely on these parameters. ATM traffic parameters relate to the characteristics of the traffic associated with an ATM connection.

Traffic parameters defined for the ATM Layer include the following:

Figure 2.25
Sample source traffic flow. AAL=ATM Adaptation Layer; PDU=protocol data unit; OAM=opera- tion, administration, and maintenance.

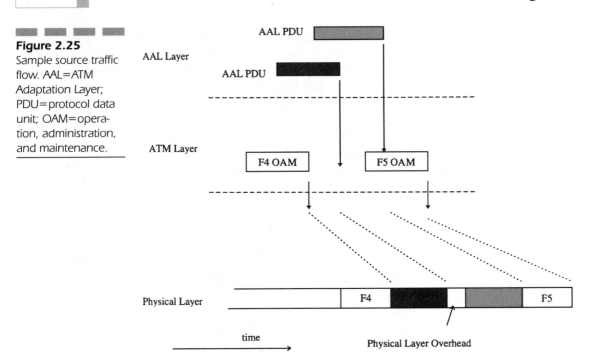

- *Peak cell rate (PCR).* The cell rate that the source may never exceed.
- *Sustainable cell rate (SCR).* The upper bound on the "average" rate of an ATM connection, over time scales that are generally long relative to those for which the PCR is defined.
- *Maximum burst size (MBS).* The maximum number of consecutive cells that can be transmitted at the peak cell rate.
- *Minimum cell rate (MCR).* A rate negotiated between the end system and the network such that the actual cell rate sent by the end system need never fall below the negotiated value. The MCR may be zero. MCR is used only with the available bit rate service category (discussed below).

To ensure that the limits set for traffic parameters are observed, the ATM Layer may perform traffic shaping functions, for example, spacing out cells transmitted over a particular connection to adhere to a PCR limit. PDUs from various traffic sources are multiplexed and shaped for delivery to the Physical Layer. The shaper is intended to provide a smoothing function to the cell flow. It ensures that at the Physical Layer, the interarrival time between two consecutive cells for a connection is greater than or equal to the inverse of the negotiated value of the PCR traffic parameter.

2.3.5   Service Categories

In general, to satisfy the needs of a range of higher-layer applications, only a subset of the traffic and QoS parameters needs to be specified for a connection. Particular combinations of these parameters are referred to as *service categories*. Thus a service category will consist of a set of QoS and source traffic parameters, and in some cases feedback mechanisms for adjustments to these parameters. The primary motivation for the specification of service categories is that it is anticipated that support of a limited set of such categories will meet the requirements of most applications. In addition, specification of a limited number of service categories can minimize the number of protocols that need to be defined.

Following is a description of each of the ATM service categories currently defined by the ATM Forum and the applications they are intended to support.

Constant Bit Rate Service Category   The constant bit rate (CBR) service category involves connections providing a fixed amount of bandwidth during the lifetime of the ATM connection. The amount of bandwidth provided is defined by the value of the PCR traffic parameter. The applicable QoS parameters include the following.

1. *Maximum cell transfer delay.* Cells that are delayed beyond the value specified by a maximum cell transfer delay are assumed to be of significantly reduced value to the application (e.g., they may be considered lost cells).

2. *Cell loss ratio.* A maximum value is specified.

3. *Cell delay variation.* A limit on the peak-to-peak variation is specified.

The CBR service category is intended to support real-time applications requiring tightly constrained delay variation (e.g., voice, video, and circuit emulation). With the CBR service category, the source may emit cells at or below the negotiated PCR and the QoS commitment will be met.

Real-Time Variable Bit Rate Service Category   The real-time variable bit rate (RT-VBR) service category is intended for real-time applications where the source rate is expected to be variable and/or bursty. The real-time nature of the application suggests a requirement for tightly constrained CTD and CDV, as would be appropriate for voice and video applications. Cells that are delayed by the network beyond a maximum CTD value specified by maxCTD are assumed to be of significantly reduced value to the application. Thus the QoS parameters specified for this category are identical to those of the CBR service category. Traffic parameters specified for this category include PCR, SCR, and MBS.

Applications that might use the RT-VBR service category would include any real-time application (including those listed above for the CBR service category) that generates traffic at a variable rate. This can allow for more efficient use of network resources through statistical multiplexing. An example would be the transmission of compressed video in a video on demand application.

Non-Real-Time Variable Bit Rate Service Category   The non-real-time VBR (NRT-VBR) service category is intended for non-real-time applications that have bursty traffic that can be characterized in terms of the same parameters used for RT-VBR, i.e., PCR, SCR, and MBS. "Non-real-time" suggests that response time is not critical to the applications. A small value for the CLR parameter is specified.

Examples of applications that might use the NRT-VBR service category are bursty applications that are sensitive to cell loss, but not to delays. The NRT-VBR service category is useful for delay-sensitive transfers because use of the SCR parameter causes the allocation of some bandwidth for the connection, as opposed to UBR and ABR (discussed below), where the traffic effectively joins a pool of bandwidth shared by other connections.

Unspecified Bit Rate Service Category   The unspecified bit rate (UBR) service category represents a "best effort" service intended for non-real-time applications that do not require tightly constrained delay or delay variation and are tolerant of cell loss. Thus, the UBR service category provides no QoS commitments for CTD, CDV, or CLR. UBR traffic may be extremely bursty. The only source traffic attribute specified for UBR connections is PCR.

Examples of applications that might make use of the UBR service category are text, data, and image applications for which "best effort" is acceptable service. Whatever the applications, they must be tolerant of potentially high cell loss.

Available Bit Rate Service Category   The available bit rate (ABR) service category can be viewed as the UBR service category enhanced for applications that are sensitive to certain QoS parameters (e.g., CTD, CDV, and CLR) and/or that require a minimum bandwidth commitment. The ABR service category will allow applications to use whatever bandwidth is available at a given point in time up to a maximum value. The ABR service category is distinguished by the incorporation of continuous feedback from the network indicating adjustment of the flow of cells into the network. The ABR flow control mechanism supports several types of feedback conveyed through ATM resource management (RM) cells, identified by the PTI code point in the cell header (see Table 2.1). This feedback mechanism permits each customer to use the maximum available bandwidth consistent with low cell loss and a notion of "fairness" in sharing the available bandwidth. No QoS commitments are provided for CDV or CTD, although cells admitted to the network are assumed not to be delayed unnecessarily.

In the ABR service category, the end system negotiates the PCR and two additional traffic parameters, the minimum cell rate (MCR) and the initial cell rate (ICR), at connection establishment. The bandwidth available to an application may vary, but does not become less than the MCR nor greater than the PCR. At any given time, an allowed cell rate (ACR) parameter is calculated (based on feedback) to determine the current bandwidth permitted to a connection. It is expected that an end system that adapts its traffic flow in accordance with the feedback will experience a low cell loss ratio and obtain a fair share of the available bandwidth.

The ABR service category is not intended to support real-time applications. Examples of applications that might use the ABR service category would include any UBR application that is willing to trade lower cost for better cell loss performance.

2.4   ATM Adaptation Layer

The ATM Adaptation Layer (AAL) lies between the ATM Layer and the next higher layer in the user plane, the control plane, and the management planes. The AAL performs functions required by the user, control, and management planes and supports the mapping between the ATM Layer and the next higher layer. It isolates the higher layers from the specific characteristics of the ATM Layer and maps the higher-layer PDUs into the information field of the ATM cell. The AAL enhances the services provided by the ATM Layer to support the functions required by the next higher layer and compensates for specific performance impairments introduced in the ATM Layer.

The AAL is further sublayered into a Segmentation and Reassembly (SAR) Sublayer and a Convergence Sublayer (CS) (see Fig. 2.26).

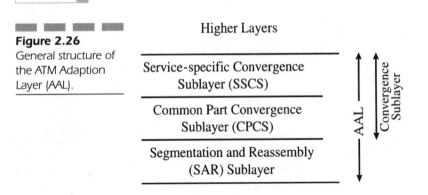

Figure 2.26

General structure of
the ATM Adaption
Layer (AAL).

1. The SAR Sublayer is the lower part of the AAL. While specifics of the SAR Sublayer vary for the individual AALs defined, its basic function is to segment higher-layer PDUs into 48-byte segments and perform reassembly at the destination. Additional SAR functions of the major AALs are discussed below.

2. The CS is the upper part of the AAL. The primary function of this sublayer is to provide any additional AAL services required to meet the higher-layer needs. In some cases, the CS Sublayer is further broken into a common part (CPCS) (providing functions useful to all users of the AAL) and a service-specific part (SSCS) (tailored to the individual user of the AAL).

As usual, AAL SAR and AAL CS entities exchange information with their peer entities to support the AAL functions.

This section briefly addresses the broadly defined ITU-T service classes whose needs the AALs are designed to meet, and discusses the major functions and protocols for the currently defined set of AALs. For more detail on AAL specifics, the reader should consult the references [I.363].

2.4.1 Service Classes

In order to minimize the number of AAL protocols, the ITU-T Recommendations define a service classification for the AAL based on the following requirements:

■ Timing relation required/not required between source and destination

TABLE 2.3

ITU-T Service
Classes

	Class A	Class B	Class C	Class D
Timing relation between source and destination	Required		Not required	
Bit rate	Constant	Variable		
Connection mode	Connection-oriented			Connectionless

■ Constant or variable bit rate

■ Connection-oriented or connectionless mode

Only a limited number of combinations of the above requirements are foreseen by the ITU. These allowed combinations result in service classes for the AAL, termed classes A, B, C, and D. Their specific requirements are indicated in Table 2.3. Examples of higher-layer services that would fall into these service classes are

■ Class A—DS1 and DS3 circuit emulation

■ Class B—packet video

■ Class C—frame relay

■ Class D—switched multimegabit data service (SMDS) [SMDS1], a connectionless data service

2.4.2 AAL Services and Protocols

Several AAL protocol types have been defined to meet the needs of the above service classes. AAL type 1 meets the needs of class A services, type 2 meets class B needs, type 3/4 meets the needs of classes C and D, and type 5 meets the needs of class C. However, there is not a strict relationship between the service classes and the protocol types, and other combinations may be used as appropriate.

AAL Type 1 The services provided by AAL type 1 are the acceptance of user information at a fixed rate and delivery of this information at the destination at the same fixed rate. This AAL also allows for the transfer of timing information between the source and the destination, the transfer of information relating to any structure associated with the user information, and the indication of lost or errored information not recovered by the AAL.

AAL type 1 has the following specific functions and associated proto-col features:

- A 3-bit sequence number is included as part of the AAL header. This allows for the detection of lost cells and the insertion of replacement cells to maintain bit count integrity for applications requiring it. This function is useful, for example, in a circuit emulation application in which framing bit positioning needs to be maintained. Typically a dummy cell of all 1s or 0s is inserted, so that although bit count integrity is maintained, individual framing errors and user data errors are likely.

- If needed, an 8-bit pointer mechanism is provided to support the structure associated with the AAL user information.

- A single-bit mechanism is provided for recovery of source clock frequency at the receiver.

- Error detection and correction mechanisms are provided via a 3-bit cyclic redundancy check (CRC) field.

In addition to the above, AAL 1 may also perform the following functions.

- Handling of cell delay variation is required to allow delivery of information to an AAL user at a constant bit rate. Buffering is used to accomplish this function.

- In order to reduce the cell payload assembly delay, the AAL PDU payload may be only partially filled with user data.

One note regarding AAL 1 is that while the protocol specification refers to an SAR Sublayer, the segmentation/reassembly function is not actually performed (or needed) by this sublayer. Rather, the function is one of accumulating regularly occurring bits or octets from the higher layers and forwarding them when sufficient information is received.

AAL Type 2   AAL type 2 allows the transport of small packets with high efficiency [Baldwin] [I.363.2]. It is specifically designed to carry low-bit-rate, variable-length, delay-sensitive packets such as would be required for packet telephony applications over an ATM network.

Briefly, AAL 2 (common part) provides what are termed logical link connections (LLCs) riding over an ATM PVC (see Fig. 2.27). These LLCs are provided by the smaller variable-length packets contained in the ATM cells. Note in the figure that each of these smaller packets contains an LLC identifier and length indicator. These LLCs are point-to-point in

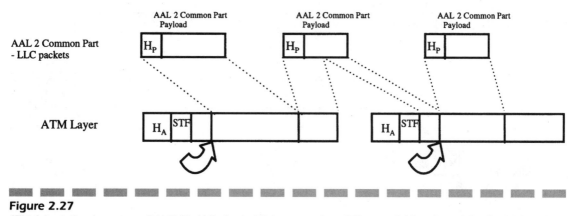

Figure 2.27
ATM Adaptation Layer type 2 (AAL 2). LLC=logical link connection. STF=start field; points to the first LLC packet in the cell. H_A=ATM cell header. H_P=AAL 2 common part packet header; contains channel identifier corresponding to the LLC and length indication representing packet length.

nature in the same manner as the PVC that carries them. There is also interest in an "LLC switch" that would provide LLC-level switching and allow multiplexing of LLCs destined for different locations so that they can be carried in the same PVC. This would provide even more bandwidth efficiency. The traffic parameters of the PVC are envisioned to be adjustable, e.g., via signaling, so that the peak cell rate can be adjusted up or down as the offered voice traffic changes. Also there is no need for an SAR Sublayer in AAL 2.

It should be noted that the principles of AAL 2 are applicable to non-ATM environments—e.g., frame relay and IP—and multiplexing protocols similar to AAL 2 for carrying low-bit-rate voice are being investigated as solutions in these areas.

Final approval of the AAL 2 Common Part is expected in 1998, and work is underway on the specification of a number of Service-Specific Convergence Sublayers.

AAL Type 3/4 AAL type 3/4 (shown in Fig. 2.28) provides for the transfer of information at variable bit rates with no timing relationship and is thus suited for data applications. Originally separate, types 3 and 4 were merged because of the similarity of their functionality. This type allows for the transfer of user data frames of from 1 to 65,535 octets. It also provides a multiplexing function, allowing multiple user data streams to be carried on a single ATM connection.

The AAL type 3/4 SAR Sublayer provides a number of features, including the following:

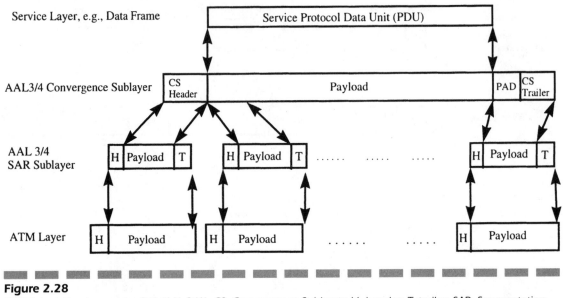

Figure 2.28
ATM Adaptation Layer type 3/4 (AAL 3/4). CS, Convergence Sublayer; H, header; T, trailer; SAR, Segmentation and Reassembly.

- A per-cell segment type indication is provided to support the segmentation and reassembly function. This 2-bit indication is used to indicate whether the 44-byte SAR payload is the beginning, continuation, or end of a message (BOM, COM, or EOM), or if the payload contains the entire message, termed a single-segment message (SSM). An EOM indication signals that reassembly can begin.

- A 4-bit per-cell sequence number is added to detect lost and/or misinserted cells.

- A 10-bit CRC field is added in the trailer of each cell to detect bit errors over the SAR PDU. Note that this allows per-cell error detection.

- A multiplexing/demultiplexing function is performed at this level, with a 10-bit multiplexing identification field used. This function provides another level of multiplexing beyond the two levels available at the ATM Layer and allows 1024 AAL connections to be multiplexed on a single ATM connection.

■ Finally, a 6-bit length indicator is encoded with the number of bytes of information that are included in the payload field. This has meaning only for EOM and SSM types, since BOM and COM types of cell will be filled (contain 44 bytes of information).

The major feature provided by the AAL type 3/4 CPCS is an indication to the receiver of buffering requirements for receiving the message.

As seen from the above, AAL type 3/4 is characterized by a rich set of services and features. However, it has generally not been widely adopted by the industry because of its complexity (many of the features can be provided at other layers) and the fact that the protocol control information used to provide these features results in a reduction in the amount of user information carried. For example, a minimum of 4 bytes of header information is used in each 48-byte SAR PDU, thus lowering the available payload by 4/48 or approximately 8 percent for a fully filled cell. Protocol control information at the CS will further reduce the available payload, although by a smaller percentage for moderately sized AAL service data units. For these reasons, the initial major application of AAL 3/4, its use for SMDS [SMDS1], has recently been defined to also operate over AAL 5 [SMDS2].

AAL Type 5   AAL type 5 (shown in Fig. 2.29) is simplified and more efficient than AAL 3/4 and is similarly mainly used for data applications. AAL type 5 has become generally preferred over AAL type 3/4 because, while not as full-function as AAL 3/4, it makes more payload available for the user data.

In contrast to the SAR Sublayer of AAL type 3/4, the SAR for AAL 5 has minimal features. One essential difference from AAL 3/4 is that the AAL 5 SAR Sublayer does not support a multiplexing/demultiplexing function. It simply accepts variable-length data units and performs the segmentation/reassembly function. No protocol fields are allocated to sequence numbering, segment type indication, or error detection at this sublayer. The sublayer does, however, use indications from the ATM Layer, passed in parameters associated with service primitives, to indicate, for example, the last segment of a message.[1] This sublayer also transparently passes other indications from adjacent layers (communicated via parameters in the service primitives), including congestion and cell loss priority.

---

[1] At the ATM Layer, this "end of message" indication is passed end-to-end via the PTI field.

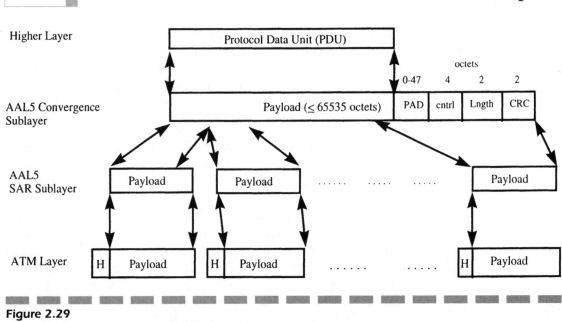

Figure 2.29
ATM Adaptation Layer type 5 (AAL 5). H, header; SAR, Segmentation and Reassembly.

Error detection is left to the AAL 5 CS Sublayer. This sublayer includes a length field to detect the loss or gain of information and a CRC field to detect errors.

2.5 References

[ANSI T1.105] ANSI T1.105, *Synchronous Optical Network (SONET)*—Basic description including multiplex structure, rates, and formats.

[ANSI T1.624] ANSI T1.624-1993, *Telecommunications*—Broadband ISDN user—network interfaces and rates and format specifications.

[ANSI/EIA/TIA-568A-95] EIA/TIA-568A *Commercial Building Telecommunications Cabling Standard,* October 1995 (ANSI/EIA/TIA-568A-95).

[Baldwin] Baldwin, John H., Behram H. Bharucha, Bharat T. Doshi, Subrahmanyam Dravida, and Sanjiv Nanda, "AAL-2—A New ATM Adaptation Layer for Small Packet Encapsulation and Multiplexing," *Bell Labs Technical Journal* (ISSN 1089-7089), Spring 1997.

[G.707] ITU-T Recommendation G.707, *Synchronous Digital Hierarchy Bit Rates,* 1993.

[G.708] ITU-T Recommendation G.708, *Network Node Interface for the Synchronous Digital Hierarchy,* 1993.

[G.709] ITU-T Recommendation G.709, *Synchronous Multiplexing Structure,* 1993.

[GR 253] Bellcore GR-253-CORE, *Synchronous Optical Network (SONET): Common Generic Criteria* (a module of TSGR, FR-440), Issue 2, December 1995.

[GR-1110] Bellcore GR-1110-CORE, *Broadband Switching System (BSS) Generic Requirements,* Issue 1, Revision 3, April 1996.

[I.356] ITU-T Recommendation I.356, *B-ISDN ATM Layer Cell Transfer Performance*—draft, September 1995.

[I.363.1] ITU-T Recommendation I.363.1, *B-ISDN ATM Adaptation Layer, AAL-1 Specification,* August 1996.

[I.363.2] ITU-T Draft Recommendation I.363.2, *B-ISDN ATM Adaptation Layer Type 2 Specification,* February 1997.

[I.363.3/4] ITU-T Recommendation I.363.3/4, *B-ISDN ATM Adaptation Layer, AAL-3/4 Specification,* August 1996.

[I.363.5] ITU-T Recommendation I.363.5, *B-ISDN ATM Adaptation Layer, AAL-5 Specification,* August 1996.

[I.371] ITU-T Recommendation I.371, *Traffic Control and Congestion Control in B-ISDN*—draft, March 1995.

[SMDS1] Bellcore TR-TSV-000773, *Local Access System Generic Requirements, Objectives, and Interfaces in Support of Switched Multi-megabit Data Service*—Issue 1, June 1991, plus Revision 1, January 1993.

[SMDS2] SIG-TS-008/1996, SMDS Interest Group, Revision 2.0, October 1996.

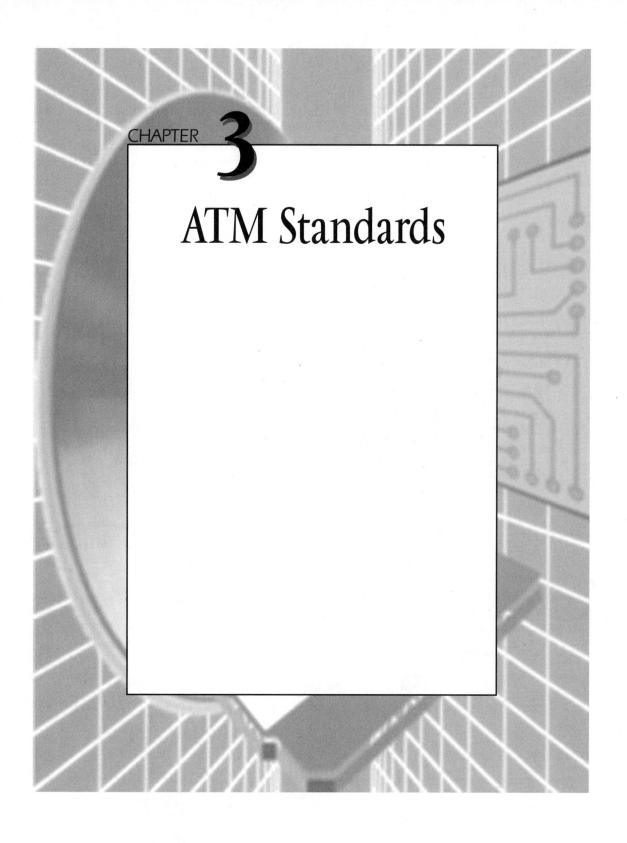

CHAPTER **3**

ATM Standards

This chapter focuses on the ATM standardization effort, which, after more than 10 years of resource-intensive research activities, has resulted in a basic set of stable standards. The chapter highlights the activities of various national and international standards bodies and industry forums and describes the key standards and specifications that have resulted from their efforts.

Figure 3.1 depicts a reference architecture for an ATM network and serves to illustrate where some of the key ATM standards apply. Four key areas of standardization are:

■ *Private User-Network Interface (UNI).* This "private UNI" applies, for example, between a workstation supporting ATM and a private ATM switch.

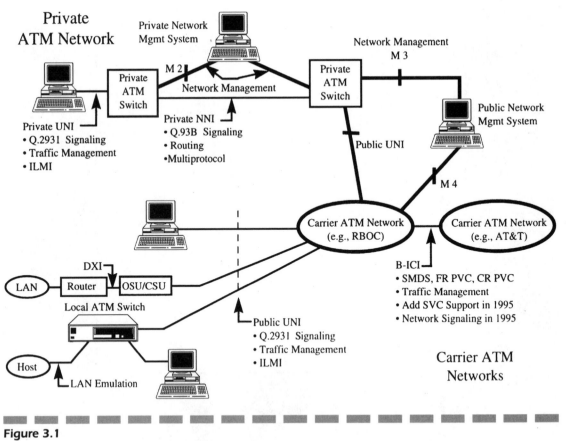

Figure 3.1

ATM Forum reference architecture and specifications.

- *Private Network-Node Interface (NNI).* This interface applies, for example, between private ATM switches needed to support the ATM infrastructure.
- *Public UNI.* This is the specification of the interface between the user and a provider of a public ATM service.
- *Public NNI.* This is the interface between ATM switches in public carrier networks. The interface is also termed the Broadband Interswitching System Interface (B-ISSI) when applied to the connection between two public switches of the same carrier and the Broadband Intercarrier Interface (B-ICI), shown in Fig. 3.1, when applied to the connection between two carriers' networks, e.g., a local exchange carrier network and an interexchange carrier's network.

The interfaces shown in Fig. 3.1 involve standards for the user plane, control plane, and management planes shown in Fig. 2-1.

3.1 Overview of ATM Standards Bodies

The following organizations have contributed to the standardization of ATM.

International Telecommunications Union—Telecommunications (ITU-T), formerly the Consultative Committee of Telephony and Telegraphy. This subagency of the United Nations aims at developing and promulgating international telecommunication standards. It has many study groups, and Study Groups XI, XIII, XV, and XVIII have made contributions to ATM standardization during the past 10 years. In particular, these study groups have developed the basic ATM protocol model and many of the key protocols associated with the model.

National bodies, such as the U.S. Alliance for Telecommunication Solutions (ATIS)—formerly the Exchange Carriers Standards Association—and, in particular, its T1 Committee. A number of T1 subcommittees have developed ATM protocol, service, signaling, performance, and operations standards. These various subcommittees are made up of major local and interexchange carriers, equipment manufacturers, and other interested parties. Generally, a standard is first developed on a national basis by technical contributions from individuals and companies. After a consensus is reached, the contribution is forwarded to the ITU-T, where international consensus is sought. National bodies, such as the T1 committee, may subsequently modify the standard to meet national requirements.

Implementors' workshops develop "implementor agreements" needed by product developers to clarify any ambiguities or lack of specificity in the standard. Examples of such workshops include the *North American ISDN User Forum,* the *ATM Forum,* and the *Frame Relay Forum.* Workshops such as the Frame Relay Forum and the ATM Forum are taking a more aggressive role than in the past, and have set out to actually develop standards themselves and/or drive the standards agenda. To date, the ATM Forum, in particular, has expedited the ATM agendas of ATIS, ITU-T, and other regional bodies in Europe and in Japan.

ATM Forum. The ATM Forum, an association of over 700 members from all segments of the computer and telecommunications industry, is an example of an implementors' workshop. Its objective is to accelerate the deployment of ATM products and services by rapid specification of standards-based interoperable implementation agreements. Formed in 1991, the group published its first interface specification, the *User-Network Interface Specification,* in 1992 and experienced a period of rapid membership growth in 1994.

Frame Relay Forum. Formed in 1991, the Frame Relay Forum is an implementors' workshop committed to the implementation of frame relay in accordance with national and international standards. In December 1994 the Frame Relay Forum, jointly with the ATM Forum, published an implementation agreement on the interworking between frame relay and ATM technologies. The implementation agreement specifies the protocol interworking functions outlined in the ITU-T I.555 standard, *Frame Relaying Bearer Service Interworking.* The agreement describes the interworking necessary to allow the interconnection of two frame relay end nodes such as frame relay access devices (FRADs) or routers via an ATM backbone network. The end devices are unaware of the use of the ATM backbone, with wide area network (WAN) switches providing the interworking functions. In April 1995, also jointly with the ATM Forum, the Frame Relay Forum published a permanent virtual circuit (PVC) service interworking agreement to allow frame relay devices to communicate with ATM devices. This interworking allows coexistence or migration of a portion of an existing frame relay network to ATM without special functionality in the end-user equipment.

Internet Engineering Task Force.   The Internet Engineering Task Force (IETF), begun in 1986, is the protocol engineering and development arm of the Internet. The IETF is a large open international community of network designers, operators, vendors, and researchers concerned with the evolution of the Internet architecture and the operation of the Internet. The IETF publishes documents termed request for comments (RFCs), some of which specify the Internet standards. Certain RFCs relate to the use of ATM in the Internet. These include the following.

RFC1626   *Default IP MTU for Use over ATM AAL5*

RFC1754   *IP over ATM Working Group's Recommendations for the ATM Forum's Multiprotocol BOF Version 1*

RFC1680   *IPng Support for ATM Services*

RFC1483   *Multiprotocol Encapsulation over ATM Adaptation Layer 5*

RFC1577   *Classical IP and ARP over ATM*

RFC1755   *ATM Signaling Support for IP over ATM*

RFC1821   *Integration of Real-Time Services in an IP-ATM Network Architecture*

Bell Communications Research (Bellcore) in specific cases adds additional requirements to standards on behalf of its clients, the regional Bell operating companies.

Additional details on key ATM standards formulated by these bodies are addressed in the remainder of this chapter. Since all ATM work, e.g., the ATM cell structure, rests on the ITU-T-related ATM standards, relevant ITU-T standards are highlighted first. The most active current ATM efforts are being carried on in working groups within the ATM Forum, and these working groups are then addressed in some detail. Since the relationship of ATM to the Internet will be a key issue going forward, IETF-related ATM standards are discussed. Finally, a listing of applicable Bellcore ATM documents is presented.

3.2   ITU-T

Table 3.1 depicts some of the key ATM standards that have been developed by the ITU-T. Figure 3.2 depicts the logical grouping of the I-series standards, the major ATM-related series. Other related series of standards, E, F, G, H, M, Q, and X, are also shown in the figure. These ITU-T standards

TABLE 3.1

Key ITU-T Standards
in Support of ATM

F.811B-ISDN	*Connection-Oriented Bearer Service*
F.812B-ISDN	*Connectionless Bearer Service*
I.113B-ISDN	*Vocabulary of Terms*
I.121R	*Broadband Aspects of ISDN* (basic principles and evolution)
I.150	*B-ISDN ATM Functional Characteristics*
I.211	*B-ISDN Service Aspects*
I.311	*B-ISDN General Network Aspects*
I.321	*B-ISDN Protocol Reference Model and Its Applications*
I.327	*B-ISDN Functional Architecture Aspects*
I.356	*B-ISDN ATM Layer Cell Transfer Performance*
I.361	*B-ISDN ATM Layer Specification*
I.362	*B-ISDN AAL Functional Description*
I.363	*B-ISDN AAL Specification*
I.365.1	*Frame Relaying Service Specific Convergence Sublayer (FR-SSCS)*
I.371	*Traffic Control and Congestion Control in B-ISDN*
I.374	*Network Capabilities to Support Multimedia*
I.413	*B-ISDN UNI*
I.432	*B-ISDN UNI Physical*
I.555	*Frame Relaying Bearer Service Interworking*
I.610	*B-ISDN OAM Principles*
I.cls	*Support for Connectionless Data Service on B-ISDN*
Q.2110	*Signaling AAL (SAAL), Service Specific Connection-Oriented Protocol*
Q.2130	*Signaling AAL (SAAL), Service Specific Coordination Function*
Q.2931	*B-ISDN Call Control*

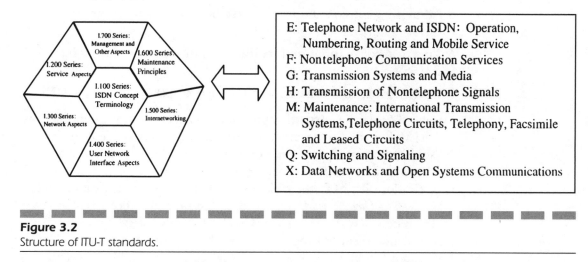

I-Series

I.700 Series: Management and Other Aspects

I.600 Series: Maintenance Principles

I.200 Series: Service Aspects

I.100 Series: ISDN Concept Terminology

I.500 Series: Internetworking

I.300 Series: Network Aspects

I.400 Series: User Network Interface Aspects

Related Series

E: Telephone Network and ISDN: Operation, Numbering, Routing and Mobile Service

F: Nontelephone Communication Services

G: Transmission Systems and Media

H: Transmission of Nontelephone Signals

M: Maintenance: International Transmission Systems, Telephone Circuits, Telephony, Facsimile and Leased Circuits

Q: Switching and Signaling

X: Data Networks and Open Systems Communications

Figure 3.2
Structure of ITU-T standards.

form the foundation upon which many of the standards of other entities are based.

3.3 ATM Forum

3.3.1 Initial Activities—ATM User-Network Interface

To ensure some level of ATM connectivity, the ATM Forum initially concentrated on the ATM UNI specification. This effort required the specification of the Physical, ATM, and ATM Adaptation Layers shown in Fig. 2-1. The user plane higher layers (e.g., the Network Layer, Transport Layer, Application Layer, etc.) are defined to operate end-to-end between user equipment and are thus beyond the scope of the ATM Forum. The control and management planes, however, involve interactions between the end user and the ATM network equipment and thus need to be defined at all layers. Thus the overall ATM UNI specification covers such diverse aspects as the following.

1. Physical media

2. The Physical Layer, ATM, and AAL protocols

3. Signaling

4. Traffic management/flow control and performance, as discussed in Chap. 2

5. Addressing considerations to specify how entities in the network can be identified

6. Treatment of point-to-multipoint and multipoint-to-multipoint traffic

7. Customer network management and carrier operations over the UNI

Table 3.2 lists the currently approved UNI specifications published by the ATM Forum. Of necessity, this work effort was carried out jointly by subject matter experts in a number of areas: Physical Layer characteristics, signaling, traffic management, and operations. The document specification shown in the table is the ATM Forum identification code.

The *ATM User-Network Interface Specification* covers both public UNIs and private UNIs shown in Fig. 3.1. The initial version of the specification did not include support for switched virtual connections (SVCs), but limited itself to permanent virtual connections (PVCs). Version 3.0 supports SVCs, and Version 3.1 was published to align with a number of international standards.

Table 3.3 depicts some of the key information exchanged at connection setup time to support SVCs. The SVC specification provides detailed formats and procedures in support of the establishment and termination of calls and the exchange of the appropriate parameters listed in Table 3.3. SVC service is of critical importance for many of the services planned to be supported with ATM. More than half of the *ATM User-Network Interface Specification,* Versions 3.0 and 3.1, deals with signaling support of the SVC capability.

The initial UNI 2.0 specification also did not deal with multipoint traffic. Versions 3.0 and 3.1 cover a basic type of multipoint service, a unidirectional point-to-multipoint traffic, primarily from an informative perspective. Much remains to be done to develop more comprehensive multipoint services that can support the sophisticated types of videocon-

TABLE 3.2		**Specification**	**Date**
Approved UNI Specifications	ATM User-Network Interface Specification, Version 2.0	af-uni-0010.000	June 1992
	ATM User-Network Interface Specification, Version 3.0	af-uni-0010.001	September 1993
	ATM User-Network Interface Specification, Version 3.1	af-uni-0010.002	1994

TABLE 3.3

Key Parameters
Exchanged at Call
Setup Time for ATM
SVC Service

Bearer capability
ATM user cell rate
Quality of Service parameter
Calling party number
Calling party subaddress
Called party/parties number/numbers
Called party/parties subaddress/subaddresses
Transit network selection
Low-layer compatibility
High-layer compatibility
Parameters in support of higher layers (e.g., adaptation functions, user-to-user signaling)

ferencing, video distribution, multimedia, distance learning, and computer-supported cooperative work for which ATM is being advocated. Table 3.4 illustrates the set of multipoint features that would ultimately be desired.

The *ATM User-Network Interface Specification* also covered an initial set of physical aspects of the UNI. For public networks, this includes SONET-based (discussed in detail in Chap. 2) fiber access (using single-mode fiber) and, because of their early availability, DS3 rate (45-Mbps) facilities. Much more discussion took place about the type of private UNI physical medium, in order to support ATM to the desktop. This UNI applies, for example, between an ATM-configured workstation and an ATM-configured hub. Initial work focused on the use of twisted pair and/or multimode fiber–based media. In addition to the actual physical medium and its characteristics, the physical portion of the *ATM User-Network Interface Specification* defines the mapping of ATM cells onto the underlying channel (e.g., the mapping to the SONET payload). Support for ATM on many additional media has been defined under the auspices of the ATM Physical Layer Working Group. These will be addressed later when the activities of that group are discussed.

Portions of the *ATM User-Network Interface Specification*, Version 4.0, have been updated and specifications published by various working groups within the ATM Forum. The specifications include *UNI Signaling*, Version 4.0, *Traffic Management*, Version 4.0, *Integrated Local Management Interface*, Version 4.0, and *Private Network-Node Interface* (*P-NNI*),

TABLE 3.4

Desired Set of
Multipoint ATM
Capabilities

Bidirectional symmetric point-to-point (BSPP) service

Bidirectional asymmetric point-to-point (BAPP) service

Unidirectional point-to-point (UPP) service

Bidirectional symmetric point-to-multipoint (BSPM) service

Bidirectional asymmetric point-to-multipoint (BAPM) service

Unidirectional point-to-multipoint (UPM) service

Bidirectional symmetric multipoint-to-multipoint (BSMM) service

Bidirectional asymmetric multipoint-to-multipoint (BAMM) service

Unidirectional multipoint-to-multipoint (UMM) service

Version 1.0. These specifications are interdependent and closely aligned, and the respective working groups closely coordinated their efforts in order to ensure compatibility among the specifications.

3.3.2 ATM Forum Working Groups

Within the ATM Forum, technical work is carried on under the auspices of the ATM Technical Committee, with the actual work performed in bodies called working groups. These working groups are established to address certain key work areas and are disbanded when work is completed. Current working groups are listed in Table 3.5. The remainder of this section discusses the charter of each of these working groups along with its major work activities, published documentation, and key issues being addressed.

Physical Layer (PHY) Working Group   Initially, Physical Layer specifications were developed within the ATM Technical Committee and published in the UNI specification. In late 1992, a separate working group, known as the PHY group, was formed to deal with Physical Layer issues. This working group develops interface specifications for transporting ATM cells at both the public and private UNI. The work includes media-dependent specifications (e.g., line encoding, modulation characteristics, and media characteristics) and media independent specifications (e.g., frame format and byte functionality). See, for example, Fig. 2.8.

Of significance are specifications to support private UNIs at 51.840 Mbps on unshielded twisted pair (UTP) Type 3 as well as at 155.52 Mbps on UTP Type 5. The motivation for UTP is to deliver ATM to the desktop without having to install fiber. Many buildings already have Type 3 UTP installed, and an increasing number of buildings have Type 5. A 25-Mbps interface over Type 3 UTP is also specified. The argument was made that the lower bandwidth is adequate to support desktop applications foreseen for the next few years.

A summary of the currently approved PHY specifications is shown in Table 3.6.

Recent activities of the PHY group include the specification of a plastic optical fiber (POF) medium for up to 155 Mbps. The forum is the first organization to agree on the specification of this medium; POF is not generally accepted by the industry, and no measurement standards currently exist. Its advantages include low cost and low installation skill, and it is being recommended in the 50-m range for home and office applications. The group has also completed a specification for inverse multiplexing for ATM (IMA) that will allow the combination of lower-speed facilities to obtain a single logical higher-speed facility for carrying ATM cells. Issues addressed include buffer design, timing and synchronization, and procedures for adding and dropping links.

TABLE 3.5

Major ATM Working Groups

PHY (Physical Layer)
B-ISDN Inter-Carrier Interface
Signaling
Network Management
Private NNI
Traffic Management
Service Aspects and Applications
LAN Emulation
Multi-Protocol over ATM
Voice and Telephony over ATM
Testing
Wireless
Residential Broadband
Security

TABLE 3.6

Approved PHY
Specifications

	Specification	Date
ATM Physical Medium Dependent Interface Specification for 155 Mbps over Twisted Pair Cable	af-phy-0015.000	September 1994
DS1 Physical Layer Specification	af-phy-0016.000	September 1994
Utopia Level 1, Version 2.01	af-phy-0017.000	March 1994
Mid-range Physical Layer Specification for Category 3 UTP	af-phy-0018.000	September 1994
6.312 Mbps UNI Specification	af-phy-0029.000	June 1995
E3 UNI	af-phy-0034.000	August 1995
Utopia Level 2, Version 1.0	af-phy-0039.000	June 1995
Physical Interface Specification for 25.6 Mbps over Twisted Pair	af-phy-0040.000	November 1995
A Cell-based Transmission Convergence Sublayer for Clear Channel Interfaces	af-phy-0043.000	January 1996
622.08 Mbps Physical Layer	af-phy-0046.000	January 1996
155.52 Mbps Physical Layer Specification for Category 3 UTP	af-phy-0047.000	
120 Ohm Addendum to ATM PMD Interface Specification for 155 Mbps over TP	af-phy-0053.000	January 1996
DS3 Physical Layer Interface Specification	af-phy-0054.000	March 1996
155 Mbps over Multi-mode Fiber, Short Wavelength Lasers, Addendum to UNI 3.1	af-phy-0062.000	July 1996
WIRE (PMD to TC Layers)	af-phy-0063.000	July 1996
E-1 Physical Layer Interface Specification	af-phy-0064.000	September 1996
155 Mbps over Plastic Optical Fiber (POF)	af-phy-0079.000	May 1997
Inverse ATM Mux	af-phy-0086.000	July 1997

Broadband Inter-Carrier Interface Working Group ATM public networks belonging to different carriers need to be interconnected in order to support end-to-end national and international ATM services. The Broadband Inter-Carrier Interface (B-ICI) Working Group defines carrier-to-carrier ATM-based multiservice interfaces to facilitate this support. The B-ICI (see Fig. 3.1) defines the point of demarcation between two networks.

The group has developed implementation specifications for PVC and SVC services between multicarrier ATM networks. Version 1.0 of the ATMF B-ICI specification supported PVC services and switched multi-megabit data service (SMDS) on ATM. This included user plane and management plane communication capabilities to support these services. Subsequent work has focused in two areas:

1. Completing enhancements of the Version 1.0 specification by adding usage metering requirements and alignment of ATM operations, administration, and maintenance (OAM) cell flows with changes made by ITU-T. The results of this work were published in Version 1.1.

2. Specifying SVC capabilities that are currently supported and defined by UNI Version 3.1. The signaling protocol is based on the ITU-T network signaling protocols, including Broadband ISDN User Part (BISUP) and the Q-series SAAL recommendations shown in Table 3.1. BISUP is a Signaling System 7 (SS7) protocol that defines the signaling messages to control connections and services. A summary of applicable standards is shown in Table 3.7. A general description of BISUP signals and messages is provided in ITU-T Recommendation Q2762. Message formats and message field codings are defined in ITU-T Recommendation Q2763, while the signaling procedures are described in ITU-T Recommendation Q2764. ANSI T1S1.3 has defined a separate specification of BISUP, consistent with the ITU specification but adding some parameters and procedures for national use. These results are documented in Version 2.0 of the B-ICI specification.

Table 3.8 summarizes the B-ICI standard at the functional level. Work on Version 3.0 of the specification has begun, with switched virtual path services and supplementary services, such as closed user groups, on the agenda.

Signaling Working Group *Signaling* is an important aspect of communication services for voice, data, video, and multimedia applications. Signaling is the process of transferring, in real time, service-related infor-

	Standard	Title
TABLE 3.7 BISUP Standards	Q2761	Functional Description of the B-ISDN User Part (B-ISUP) of Signalling System No. 7
	Q2762	General Functions of Messages and Signals of the B-ISDN User Part (B-ISUP) of Signalling System No. 7
	Q2763	Signalling System No. 7 B-ISDN User Part (B-ISUP)—Formats and Codes
	Q2764	Signalling System No. 7 B-ISDN User Part (B-ISUP)—Basic Call Procedures

TABLE 3.8

Highlights of the
B-ICI Specifications

The specification includes support of PVC and SVC cell relay service.

The specification includes support of intercarrier services such as circuit emulation service (CES), frame relay service, and SMDS.

The Physical Layer specification includes the SONET rates as well as DS3.

The ATM Layer specification is common to all the B-ICI Physical Layers.

The service-specific functions above the ATM Layer include the AAL and network interworking.

The specification includes traffic management and network performance.

The specification includes operations and maintenance.

mation between the user and the network and among network entities to establish end-to-end communications. Signaling enables the user to specify the details of the connection via the exchange of the values of parameters shown previously in Table 3.3.

It is important to note that signaling takes place at two points.

1. *Over the UNI.* The *ATM User-Network Interface Specification,* Versions 3.0 and 3.1, covers public UNI as well as private UNI signaling; the same protocols are used on both interfaces. The specifications also cover point-to-multipoint signaling and signaling for new service classes. The specifications are based on the ITU-T Q2931 standard. The Signaling ATM Adaptation Layer (SAAL) protocol and procedures provide reliable delivery of Q2931 signaling messages and define how to transfer the signaling information for call and connection control within the ATM Layer on virtual channels used for signaling. The SAAL resides in the control plane and, like other AALs discussed in Chap. 2, is made up of a common part, which represents the functionality common to a connection-oriented, variable bit rate information transfer, and a service-specific part, which identifies the protocol and procedures associated with the signaling needs at the UNI. The common part protocol (based on the AAL5 common part) provides unassured information transfer and a mechanism for detecting corruption of information carried in the SAAL frames; the service-specific part provides recovery. As indicated in Table 3.1, the service-specific part is further subdivided into a service-specific coordination function (SSCF or Q2130) and a service-specific connection-oriented protocol (SSCOP or Q2110). The SSCOP is used to transfer variable-length service data units (SDUs) between users and provides for the recovery of lost or corrupted SDUs. The SSCF maps the services of SSCOP to the needs of the SSCF user.

2. *Between network elements.* For signaling between network elements, there are some choices for implementation. In public networks, signaling between network elements is based on the SS7 protocol standards and use of the SS7 network. This approach allows the continued support of transaction services such as 800, credit card calls, etc., in an ATM network. An alternative signaling approach for public networks (possibly used in the early deployment stages where the number of switches is small) could involve direct switch-to-switch signaling links. However, in this case, the capability to support transaction-based services would be lost.

For interfaces between network elements in private networks, the use of an appropriate modification of Q2931 over SAAL over ATM is an alternative to the SS7 protocols. This is discussed further in conjunction with the P-NNI Working Group, whose work is covered later in this section.

The *UNI Signaling* 4.0 specification, published in July 1996, adds a number of useful features to the basic call setup capability of *UNI* Version 3.1. These features include support for available bit rate (ABR) service, enhanced Quality of Service support, and proxy signaling to support video on demand. Of necessity, the Signaling Working Group works closely with other groups, particularly the Traffic Management and P-NNI Working Groups, to align work activities.

Network Management Working Group   The Network Management (NM) Working Group develops specifications, recommendations, and guidelines for network and systems management. Specific work areas include the specification of a number of management interfaces—public/private, network operator/customer—as shown in Fig. 3.1.

Layer Management capabilities can be implemented using either the specialized OAM cells discussed in Chap. 2 or a higher-layer management application known as the Integrated Local Management Interface[1] (ILMI). The ILMI 4.0 specification, published in September 1996, enables two adjacent ATM devices to automatically configure the operating parameters of the common ATM link between them. In addition, the ILMI 4.0 specification provides communication across the ATM link, enabling both sides to verify subscription parameters, the configuration of newer traffic services such as ABR, signaling version identification, and address registration information. Table 3.9 depicts some of the key functions supported via the ILMI.

Management plane capabilities are supported via higher-layer protocols such as the Open Systems Interconnection (OSI) Common Management Information Service Element/Common Management Interface

---

[1]This interface was previously termed the *Interim* Local Management Interface.

TABLE 3.9

Integrated Local Management Interface (ILMI) Functions

Physical Layer	Interface index
	Interface address
	Transmission type
	Media type
	Operational status
ATM Layer	Maximum number of virtual path connections (VPCs)
	Maximum number of virtual channel connections (VCCs)
	VPI/VCI address width
	Number of configured VPCs
	Number of configured VCCs
	Port type
ATM Layer statistics	ATM cells received
	ATM cells dropped on receive side
	ATM cells transmitted
VP connection	VPI value
	Shaping traffic descriptor
	Policing traffic descriptor
	Operational status
	Quality of Service (QoS) category
VC connection	VPI/VCI value
	Shaping traffic descriptor
	Policing traffic descriptor
	Operational status
	QoS category

Protocol (CMISE/CMIP) or Simple Network Management Protocol (SNMP). Preliminary management capabilities are addressed in a number of documents, including the ATM Forum's *ATM User-Network Interface Specification* and *B-ISDN InterCarrier Interface (B-ICI) Specification*, the ITU-T I.610 specification, and Bellcore documents (see Sec. 3.3.4). The ATIS T1S1.5 subcommittee has also been very active in developing OAM Layer Management functions. Initial work has been directed at PVC services; more recent work has been directed at SVC services.

As indicated in Table 3.10, the NM Working Group has primarily focused on the specification of the M4 interface for management system interfaces to ATM public network elements (see Fig. 3.1). The ATM Forum originally approved the *M4 Interface Requirements and Logical MIB* specification. This specification includes the requirements for the network management functionality provided over the M4 interface, along with an object-based definition of the interface, specified in a protocol-independent manner to allow the management of equipment from different suppliers. This specification can be used to specify a CMIP, an SNMP, or even a proprietary interface. The group has also specified a Customer Network Management (CNM) interface to allow customers to manage public network ATM services via an automated interface. Network management specifications are summarized in Table 3.10.

TABLE 3.10

Approved NM Specifications

	Specification	Date
Customer Network Management (CNM) for ATM Public Network Service	af-nm-0019.000	October 1994
M4 Interface Requirements and Logical MIB	af-nm-0020.000	October 1994
CMIP Specification for the M4 Interface	af-nm-0027.000	September 1995
M4 Public Network View	af-nm-0058.000	March 1996
M4 Network Element View	af-nm-0071.000	January 1997
Circuit Emulation Service Interworking Requirements, Logical and CMIP MIB	af-nm-0072.000	January 1997
M4 Network View CMIP MIB Specification, Version 1.0	af-nm-0073.000	January 1997
M4 Network View Requirements and Logical MIB Addendum	af-nm-0074.000	January 1997

Private Network-Node Interface (P-NNI) Working Group This working group defines private network-node interfaces to allow the interconnection of multisupplier ATM networks. The initial work of this group was focused on an interim solution, driven by time-to-market concerns, termed the Interim Interswitch Signaling Protocol (IISP). That specification was based on UNI Version 3.1. The IISP uses static, fixed routing with next-hop routing tables and exchanges no link-state information. P-NNI Version 1.0 was subsequently specified to support both dynamic hierarchical state-of-the-art routing between switches and clusters of switches, and signaling between the switches. The routing protocol, based on link-state routing techniques, is expected to scale to large worldwide ATM networks. The signaling protocol is based on the UNI signaling specification (see the discussion in the Signaling Working Group section above) and has mechanisms added to support source routing and alternate routing in case of connection setup failure. The P-NNI protocols may also be used, where desired, for public network interfaces. Approved specifications are listed in Table 3.11.

Traffic Management Working Group The Traffic Management (TM) Working Group defines procedures and parameters related to traffic control and Quality of Service. As discussed previously, traffic management functions protect the network and end systems from congestion and assist in the efficient use of network resources. Approved specifications from this group are listed in Table 3.12.

A major activity of this group was the definition of the ABR service and associated traffic controls to support the service. ABR service aims at making optimal use of the available bandwidth by dynamically and fairly allocating the bandwidth on an as-needed basis to all existing active connections. This traffic class can be used in an ATM LAN and in the interconnection of ATM LAN environments. The group also identified ATM Layer performance parameters needed to characterize an ATM connection.

The *Traffic Management,* Version 4.0 specification now includes a service architecture, definitions for QoS parameters, and the specification

TABLE 3.11		Specification	Date
Approved P-NNI Specifications	*Interim Interswitch Signaling Protocol*	af-pnni-0026.000	December 1994
	P-NNI Version 1.0	af-pnni-0055.000	March 1996
	P-NNI Version 1.0 Addendum (soft PVC MIB)	af-pnni-0066.000	September 1996

TABLE 3.12

Approved TM Specifications

	Specification	Date
Initial work published in UNI 3.1	af-uni-0010.002	
Traffic Management, Version 4.0	af-tm-0056.000	April 1996
Traffic Management ABR Addendum	af-tm-0077.000	January 1997

TABLE 3.13

Approved Service Aspects and Applications Specifications

	Specification	Date
Frame UNI	af-saa-0031.000	September 1995
Circuit Emulation	af-saa-0032.000	September 1995
Native ATM Services: Semantic Description	af-saa-0048.000	February 1996
Audio/Visual Multimedia Services: Video on Demand Specification, Version 1.0	af-saa-0049.000	January 1996
Audio/Visual Multimedia Services: Video on Demand Specification, Version 1.1	af-saa-0049.001	March 1997
ATM Names Service	af-saa-0069.000	November 1996

of an ABR flow control protocol that can be used to provide bandwidth on demand on an end-to-end basis.

Service Aspects and Applications Working Group   The Service Aspects and Applications (SAA) Working Group provides specifications to enable new and existing applications to use an ATM network. The group also provides mapping and interworking specifications for existing LAN and WAN services over ATM, e.g., frame relay/ATM interworking. Specifications published by this group are shown in Table 3.13.

A major effort of this group has been the definition of the audiovisual multimedia service. This work entails the transport of MPEG-2 over an appropriate ATM AAL for multimedia conferencing and video distribution. The effort facilitates the definition of multimedia conferencing and video distribution solutions that are optimized for the ATM cell-based environment, taking advantage of the bandwidth capabilities of ATM as well as of its traffic management capabilities.

LAN Emulation (LANE) Working Group   This working group develops specifications in support of interconnecting existing LANs (Ethernet and token ring) over an ATM backbone. Currently most enterprise data traffic originates on IEEE 802.3 and 802.5 LANs, and end users will continue using LAN-based applications. LANE is an ATM service, supported by software in the end system, that emulates the services of existing LANs across an ATM network. LANE allows the higher layers in the end system to view the network as a conventional LAN. Essentially, LANE provides Media Access Control (MAC) Layer emulation, allowing migration to ATM technology. Its specifications also support interconnecting ATM-attached servers and workstations to each other and to legacy LANs.

The protocol architecture supporting LANE is shown in Fig. 3.3. LANE defines a number of functional components, including the following:

LAN emulation client (LEC).   An end station or bridge running LANE is referred to as a LAN emulation client.

LAN emulation service (LE Service).   A number of functional entities are defined to provide the LANE emulation service; these include

- *LAN Emulation Server (LES).*   This provides a facility for registering and resolving MAC addresses.
- *Broadcast and Unknown Server (BUS).*   This handles data sent to the broadcast MAC address (FFFFFFFFFFFF), multicast traffic, and initial unicast frames before address resolution.
- *LAN Emulation Configuration Server (LECS).*   This assigns individual LECs to a particular emulated LAN.
- *LAN Emulation User-to-Network Interface (LUNI).*

Currently available specifications are shown in Table 3.14. Version 1.0 did not provide a method for redundant servers to back up LANE devices, resulting in the potential for single points of failure. Version 2.0 will provide this feature.

Multi-Protocol over ATM Working Group   The Multi-Protocol over ATM (MPOA) Working Group develops protocols and mechanisms to enable layer 3 protocols to operate over an ATM network. This work will support native ATM hosts and hosts connected to legacy networks and via LANE. The initial specification was published in July 1997.

Voice and Telephony over ATM (VTOA) Working Group   The ATM Forum started work on voice transport in 1993, and in early 1995 the VTOA Working Group published its first document, addressing both unstructured and structured circuit emulation specifications. Unstructured circuit emulation maps an entire T1 (1.544-Mbps) circuit to a single ATM virtual circuit (VC), thus limiting it to point-to-point applications.

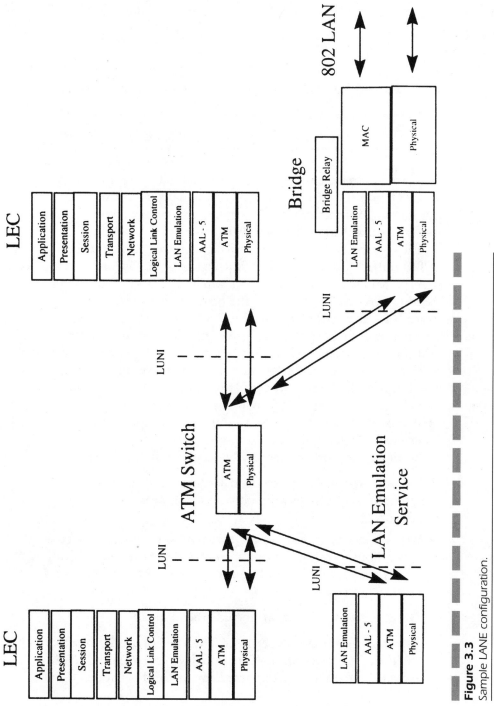

Figure 3.3
Sample LANE configuration.

TABLE 3.14

Approved LANE
Specifications

	Specification	Date
LAN Emulation over ATM, Version 1.0	af-lane-0021.000	January 1995
LAN Emulation Client Management Specification	af-lane-0038.000	September 1995
LANE Version 1.0 Addendum	af-lane-0050.000	December 1995
LANE Servers Management Specification, Version 1.0	af-lane-0057.000	March 1996
LANE Version 2.0 LUNI Interface	af-lane-0084.000	July 1997

Structured circuit emulation allows switches to map individual 64-kbps circuits in a T1 line to ATM VCs, and it can be used for point-to-multipoint connections. An update to the circuit emulation specification was published in January 1997.

Each specification requires voice to be treated as constant bit rate (CBR) traffic. A problem with CBR traffic is that it forces customers to reserve bandwidth for voice even when no information is being sent. Sending voice as VBR traffic is the obvious alternative; however, VBR for voice has not yet been standardized. Silence suppression and voice compression will be part of a new AAL specification, AAL 2, that will provide more efficient use of bandwidth. AAL 2, discussed in Chap. 2, defines a VBR service for low-bit-rate voice traffic and is expected to be used initially between a wireless base station and a mobile switch center. When fully defined, AAL 2 will support a signal processing layer that will provide silence suppression and voice compression. Some vendors offer pre-standard equipment.

Work in this group has generally been split into two areas:

1. VTOA trunking for narrowband services, targeted primarily at applications in private voice networks and potentially at public networks

2. VTOA legacy voice services at a native ATM terminal, targeted at applications in private and public networks where interworking and interoperation of ATM and non-ATM networks and services for voice are necessary.

Other VTOA specifications include the following:

■ *Trunking for Narrowband Services.* This specification is based on the use of an interworking function (IWF) between the ATM network and each interconnected narrowband network.

■ *Voice and Telephony over ATM to the Desktop Specification.* This document specifies the functions required to provide voice and telephony services over ATM to the desktop. It describes the functions of the IWF and a native ATM terminal. This version covers only the transport of a single 64-kbit/s A-law or mu-law encoded voiceband signal.

Testing Working Group The Testing Working Group provides test specifications for the protocols defined at the various layers of the ATM protocol architecture. Specifications of this group are shown in Table 3.15.

Wireless ATM (WATM) Working Group This working group is developing specifications to facilitate the use of ATM technology for a range of wireless network access scenarios, including both terminal mobility and radio access. This working group's importance is due to the fact that new high-speed-capable spectrum is becoming available, and this new spectrum can support multiple services (voice, video, data). The characteristics of the new spectrum call for a short packet interface, and this group has been formed to ensure that ATM can provide this interface.

Residential Broadband Working Group The Residential Broadband (RBB) Working Group was formed in April 1995 to define a complete end-to-end ATM system both to and from the home and within the home. The purpose of this effort is to specify common interfaces that vendors can build to, thereby accelerating the use of ATM within the residential broadband market.

Many interactive services will be supported over a residential broadband network, and each residence may have more than one device—PCs, set-top boxes, etc.—accessing these services. The RBB group has defined a reference configuration with three parts. First, there is a core ATM network to which all of the services, e.g., video servers, are connected. The current ATM Forum specifications are sufficient for the interfaces used in this network. Second, an access network connects to the core network and delivers the services to the various homes. Third, a UNI is presented by the access network to each home. A network terminator (NT) terminates the access network and separates the access network from the home network. Work is in progress in the RBB Working Group to define these various interfaces and adapters.

TABLE 3.15

Approved Testing
Specifications

	Specification	Date
Introduction to ATM Forum Test Specifications	af-test-0022.000	December 1994
PICS Proforma for the DS3 Physical Layer Interface	af-test-0023.000	September 1994
PICS Proforma for the SONET STS-3c Physical Layer Interface	af-test-0024.000	September 1994
PICS Proforma for the 100 Mbps Multimode Fiber Physical Layer Interface	af-test-0025.000	September 1994
PICS Proforma for the ATM Layer (UNI 3.0)	af-test-0028.000	April 1995
Conformance Abstract Test Suite for the ATM Layer for Intermediate Systems (UNI 3.0)	af-test-0030.000	September 1995
Interoperability Test Suite for the ATM Layer (UNI 3.0)	af-test-0035.000	April 1995
Interoperability Test Suites for Physical Layer: DS-3, STS-3c, 100 Mbps MMF (TAXI)	af-test-0036.000	April 1995
PICS for DS-1 Physical Layer	af-test-0037.000	April 1995
Conformance Abstract Test Suite for the ATM Layer (End Systems) (UNI 3.0)	af-test-0041.000	January 1996
PICS for AAL5 (ITU specification)	af-test-0042.000	January 1996
PICS Proforma for the 51.84 Mbps Mid-Range PHY Layer Interface	af-test-0044.000	January 1996
Conformance Abstract Test Suite for the ATM Layer of Intermediate Systems (UNI 3.1)	af-test-0045.000	January 1996
PICS for the 25.6 Mbps over Twisted Pair Cable (UTP-3) Physical Layer	af-test-0051.000	March 1996
PICS for ATM Layer (UNI 3.1)	af-test-0059.000	July 1996
Conformance Abstract Test Suite for the UNI 3.1 ATM Layer of End Systems	af-test-0060.000	June 1996
Conformance Abstract Test Suite of the SSCOP for UNI 3.1	af-test-0067.000	September 1996
PICS for the 155 Mbps over Twisted Pair Cable (UTP-5/STP-5) Physical Layer	af-test-0070.000	November 1996
PNNI Version 1.0 Errata and PICS	af-pnni-0081.000	July 1997
PICS for Direct Mapped DS3	af-test-0082.000	July 1997
Conformance Abstract Test Suite for Signalling (UNI 3.1) for the Network Site	af-test-0090.000	September 1997

Several other standards bodies and working groups are also working on various aspects of the residential broadband network, and the RBB Working Group maintains a close and regular liaison with many of these groups. For example, as mentioned previously, this group is working closely with the PHY Working Group to define the POF specification.

Security Working Group The Security Working Group was formed in December 1995 and its planned phase 1 specification is scheduled for completion in 1998. The specification will address confidentiality and authentication issues, with strong authentication being the primary focus of this initial phase. Security is being addressed first in the user plane, with some control plane security issues also being addressed. Management plane security will likely be a topic for phase 2 work of this group.

3.3.3 Additional ATM Forum Specifications

Data Exchange Interface (DXI) The ATM DXI version 1.0 specification allows data terminal equipment (DTE) (such as a router) and data communications equipment (DCE) [such as an ATM Channel Service Unit (CSU)] to cooperate in providing a UNI for ATM networks. This is expected to expedite introduction of ATM services in user environments, since users will not have to completely replace existing equipment. For example, there are situations in which the user may want to retain an existing (and programmable) router. With the DXI approach, the user loads appropriate software (probably supplied by the DXI CSU vendor) into the router and adds in the DXI-configured CSU, enabling access to an ATM network. DXI-configured CSUs were also used in SMDS, although these CSUs implemented the SMDS protocols, not the ATM protocols. In SMDS a number of router vendors took this approach; others implemented the entire SMDS stack. Over time the importance of DXI can naturally be expected to decrease.

Anchorage Accord Specifications In early 1996, the ATM Forum saw the need to introduce a structure for all the specifications produced by the organization. This was based largely on industry feedback that the rate of publication of ATM specifications was such that the industry was having a difficult time determining the importance of the various specifications. The result of this activity was the Anchorage Accord,

TABLE 3.16

ATM Forum Foundation Specifications

User Network Interface (UNI), Version 3.1

BICI, Version 2.0, providing SVC capabilities between switched networks based on UNI 3.1 capabilities

Integrated Local Management Interface (ILMI), Version 4.0, allowing verification of subscription parameters and address registration

Network management specifications, providing management capabilities for M3 and M4 interfaces and MIBs

Physical Layer specifications at the UNI, allowing use of existing cabling structures as well as opportunities to upgrade, and also supporting transmission hierarchies around the world, including Europe and Japan (e.g., DS1; J2; E1; DS3; E3; 25-Mbps and 50-Mbps UTP 3; 155-Mbps UTP 5; 100 Mbps, 155 Mbps, and 622 Mbps using multimode fiber; 155 Mbps and 622 Mbps using single-mode fiber; fiber driven by LEDs and lasers, etc.).

Interim Interswitch Signaling Protocol (ISSP), supporting statically configured ATM corporate/enterprise networks

Private Network-Node Interface (P-NNI), Version 1.0, enabling switch-switch private/enterprise networks and introducing P-NNI routing and symmetrical signaling based on *Signaling*, Version 4.0 to support dynamic hierarchical state-of-the-art routing

Signaling, Version 4.0, with signaling capabilities for ABR service, new traffic descriptors, and point-to-multipoint connections

Traffic Management, Version 4.0, defining an architecture for the ATM Layer service categories and also defining a set of functions and procedures to manage and control traffic and congestion, and introducing the ABR flow control protocol.

which defined certain specifications as being *foundation* specifications (shown in Table 3.16), needed to build an ATM infrastructure, and other specifications as being *expanded feature* specifications, needed to enable ATM to support multiple services. It is envisioned that the foundation specifications will be rather stable, with revisions to correct problems discovered during implementation and maintain ITU-T alignment. Expanded feature specifications generally include service or service interworking specifications, such as multimedia services, LANE, and frame relay/ATM service interworking. Existing expanded feature specifications may change more rapidly than the foundation specifications, and new expanded feature specifications will be added as appropriate.

3.3.4 Bellcore Documents

Figure 3.4 depicts the set of documentation developed by Bellcore on behalf of its clients, primarily the regional Bell operating companies, in

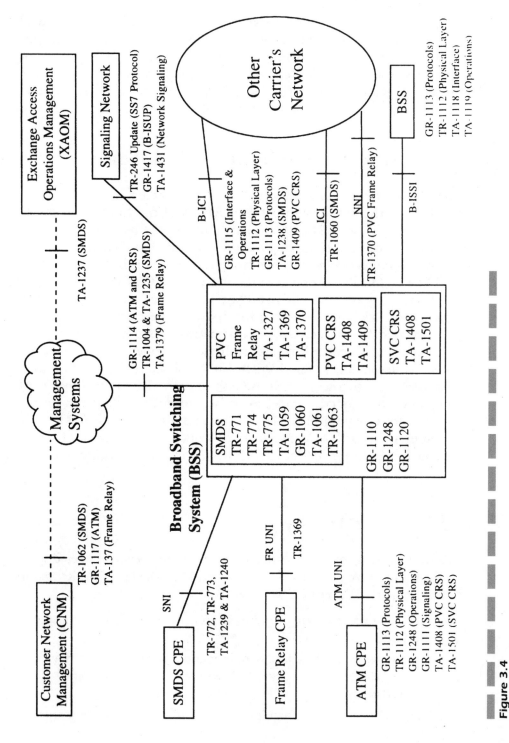

Figure 3.4

Bellcore ATM specifications for Broadband Switching Systems (BSSs).

support of ATM and fast packet services. Many of these technical requirements (TRs), technical advisories (TA), and generic requirements (GRs) are based on published standards, but discuss relevant subsets of features. Some of these Bellcore documents have, in fact, driven the industry standards, being first published as industry requirements and then finding their way into the standards.

3.3.5 Internet Engineering Task Force

As previously mentioned, the IETF has published a number of RFCs addressing the relationship between ATM and the Internet protocols. Key among these is RFC 1577, *Classical IP and ARP over ATM*. RFC 1577 follows the classical IP routing model, in which IP packets from subnetwork to subnetwork must flow through routers. This is the case in RFC 1577 for ATM subnetworks even though a direct ATM path may be possible between the end stations. Thus, nothing new is required for IP packet forwarding (in hosts or routers), and the classical hop-by-hop routing paradigm is retained. Intra-subnetwork communication requires IP entities to discover link-level addresses of target IP entities via an Address Resolution Protocol (ARP). Unlike broadcast-capable link technologies (e.g., Ethernet), which use peer-to-peer ARP mechanisms, ATM requires a client-server model, as native broadcast/multicast is unsupported. To accomplish this, RFC 1577 defines client-server ATMARP procedures, with each ATM *logical* subnetwork having its own ATMARP server. Procedures are defined for clients to register with this server.

3.4 Summary

A large number of ATM standards have already been completed, published, and implemented, making basic ATM services possible. Related to these standards, the critical issues are

- How soon vendors bring equipment to market that implements the standards
- To what degree there will be vendor compliance in terms of how much of the standard is actually supported (implementation of nonoverlapping subsets will lead to lack of interoperability)
- The cost-effectiveness of equipment implementing these standards

ATM products are currently available from over 200 suppliers, including network interface card (NIC) suppliers, software providers, switch product suppliers for public/WAN and private network applications, and chip set suppliers. ATM PVC services are available from several carriers, and SVC services are starting to be deployed.

Enough standards have been published to enable the basic technology to see commercial introduction. In addition, standards are available to enable ATM to support ATM/frame relay service interworking, provide point-to-point transport of compressed digital video and multimedia, and coexist with legacy protocols (such as IP) and legacy LANs (such as Ethernet). Current challenges are to move from early proprietary products to nonproprietary system and network implementations and to minimize differences between the ATM Forum specifications and ITU-T recommendations.

ATM Switch Architecture

4.1 Functional View

For purposes of understanding the workings of an ATM switch, it is convenient to discuss the switch in terms of its major functional components. This functional view also aids in discussion of the design alternatives for switch implementations. A number of functional models have been proposed in the industry, and we make use of these with updates to incorporate recent developments such as routing [Toba] [Chen] [Bell] [Hels].[1]

Figure 4.1 shows a basic functional model, with the following major components:

- *Input functions* concerned with terminating transmission lines and extracting ATM cells

- *Output functions* concerned with preparing the outgoing signals for transmission to the next switch (or to the end user)

- *Cell switching functions* to support the switching of ATM cells based on incoming/outgoing virtual path identifier/virtual channel identifier (VPI/VCI) values

---

[1]To quote one source is plagiarism; to quote many is research.

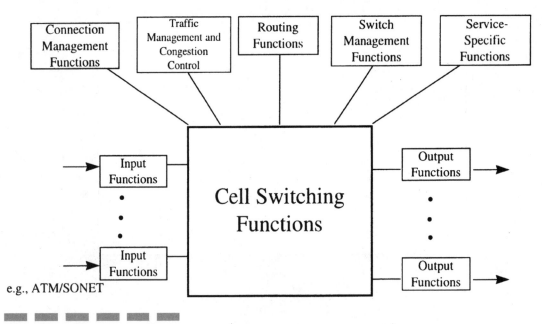

Figure 4.1
ATM switch functional architecture.

- *Connection management functions* concerned with handling connection/disconnection requests (and the subsequent acceptance or rejection of the requests), assigning and maintaining connection parameters such as VPI/VCI, and recovering from connection failures

- *Traffic management and congestion control functions* concerned with detecting the onset of congestion and invoking actions to minimize the effect of the congestion

- *Routing functions* concerned with the use of topology information to determine routes through the network (in some cases this may involve the exchange of network topology information among ATM switches)

- *Switch management functions* concerned with operations and management aspects of the ATM switch, including configuration management, fault management, performance management, accounting management, and security management

- *Service-specific functions,* which include interworking functions required to adapt non-ATM-based services to be switched and transported by ATM-based networks (frame relay service being one example)

Note that some of the above functions, such as input and output, switch management, and cell switching, are within the scope of a single switch, whereas others, such as connection management, traffic management and congestion control, and routing, are distributed in nature and require cooperation with other switches within the network. There is also some inevitable arbitrariness and overlap in the functional breakdown noted above. For example, a major part of the traffic management and congestion control function is the determination of whether sufficient resources are available to satisfy a connection request without placing the switch/network in danger of congestion (this function is termed connection admission control, or CAC). Logically, the CAC function could also be thought of as part of the connection management function.

Note also that this functional view is not meant to imply a specific physical architecture for an ATM switch. Part of the switch design process is, in fact, the assignment of the above functions to a particular hardware element or elements. For example, one common design decision for many functions is choosing a centralized or distributed implementation; the tradeoff for the efficiency gained by physically distributing particular functions to a number of hardware elements rather than locating the function at a single central element is the added complexity associated with distributed processing.

The remainder of this chapter discusses the above functions and issues affecting their physical implementation and placement within the ATM switch. Sample industry switch designs will be highlighted where appropriate to illustrate these design choices, particularly those associated with the cell switching functions, a key element of the ATM switch architecture. Detailed designs for specific switches, the Telematics NCX-1E6 and the Ascend B-STDX 9000, are covered in later chapters.

4.2 Input Functions

Examples of input functions include the SONET functions discussed in Sec. 2.2 and shown in the lower portion of Fig. 2.8. Typically, input functions are physically located on an input module, shown in Fig. 4.2, and include the following:

- The input signals are converted from the optical signal received over a fiber-optic facility to an electrical signal (O/E conversion) for subsequent processing via electronic circuitry.

- Timing for the switch is derived from the electrical signal, if required, and SONET Section Layer functions, including framing and descrambling, are performed.

- SONET section, line, and path overhead are processed, and information relating to the SONET Layer operations, administration, and maintenance (OAM) functions (e.g., occurrence of parity errors) is passed to the switch management function.

- Cell delineation, i.e., determining the location of the cell boundaries in the SONET payload, is performed as discussed in Sec. 2.2.5, based on verification of a valid header error control (HEC) byte.

- After the ATM cells are extracted from the SONET payload, idle and unassigned cells are dropped and the header of valid cells is checked via the HEC byte. Header error control operates in a number of modes, with single-bit error detection and correction the default mode. Once an error is detected, an alternative mode with multiple-bit error detection and no error correction is entered. After the detection and correction of a single-bit error, the alternative mode is employed until a cell with no errors is received, at which time the single-bit error correction mode is reentered. In

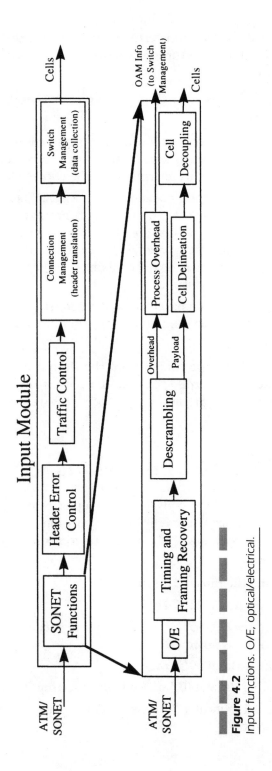

Figure 4.2

Input functions. O/E, optical/electrical.

the error detection mode, errored cells are discarded and notification of the error is forwarded to the switch management function.

Input functions may closely interact with other functions. For example,

- For an established connection, the user's traffic is monitored (as part of the traffic management and congestion control function, discussed in Sec. 4.6) to determine its adherence to the traffic contract.

- As part of connection management, a cell may be prepared for internal routing by the cell switching functions through a physical cell switching fabric. This typically involves attaching a locally significant routing tag based on the output and input ports, cell priority, and whether the cell contains signaling, user, or management information.

- The collection of management and accounting data, part of the switch management function, is accomplished in cooperation with the input function.

Because of this close interaction, the design of a switch generally involves the physical collocation of the above functions, along with the input functions, on physical input modules as shown in Fig. 4.2.

4.3   Output Functions

Output functions are considerably simpler than the input functions and involve preparing the cells for transmission on outgoing facilities. In the case of SONET, Fig. 4.3 shows the functions performed. They include the following:

- Removing the internal routing tag
- Generating a new HEC field
- Mapping the cells to the SONET payload, including the insertion of idle or unassigned cells
- Generating the SONET overhead
- Scrambling the SONET payload and inserting the framing bytes
- Electrical-to-optical (E/O) conversion for transmission on fiber-optic facilities

Output Module

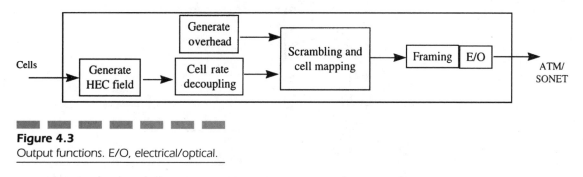

Figure 4.3
Output functions. E/O, electrical/optical.

4.4   Cell Switching Functions

ATM switches differ from conventional switches because of the high-speed nature of the input and output interfaces discussed above and the resulting high rates associated with the hardware fabric performing the switching functions (currently in the 100 Gbps range). Cell switching functions are concerned with the exchange (switching) of cells among input/output (I/O) switch modules. This switching function can be performed on both user cells and the OAM cells discussed in Sec. 2.3.1. The basis for this exchange of cells within the ATM switch modules is information stored in the switch connection table. This information is constructed and maintained as part of the connection management function, discussed in Sec. 4.5. Connection table information is reflected in the routing tag assigned to the cell as part of the input module function, discussed above. Cell switching functions also need to provide support for multicast and broadcast capabilities, congestion monitoring and notification, and support of the various traffic classes discussed in Sec. 2.3.5.

Cell switching functions are not standardized, and ATM switch manufacturers use a wide variety of techniques to differentiate their products. This section discusses the three major techniques used in implementing the cell switching functions—shared memory, shared medium, and space division—and the tradeoffs associated with each. Criteria generally applied in the evaluation of these techniques include the following:

■ *Blocking characteristics,* i.e., the likelihood that packets destined for *distinct* output ports cannot be transferred as a result of contention among shared resources within the switch fabric. For example,

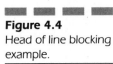

Figure 4.4
Head of line blocking
example.

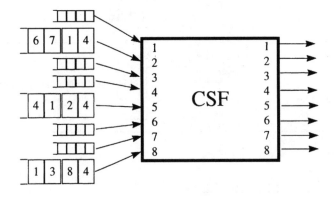

- Only one of the three input buffers with cells for output port 4 will be able to transmit
- Others will send no cell, even though cells in the queue are destined for idle ports

resource contention occurs whenever more than one packet has to access the same internal link.

■ *Cell buffering characteristics.* Cell buffering is a major requirement in implementing the cell switching functions. This buffering accounts for contention at the output ports, which occurs whenever more than one input port has traffic destined for a single output port. Physical placement of the buffers within the cell switching fabric is a major design issue, with input port, output port, and internal fabric buffers being the major choices. Storage at the output is becoming the preferred choice; if cells are stored at the input port or internal to the switch fabric, those cells queued for one port can slow down cells destined for other ports. One example of such a problem is termed head of line (HOL) blocking, shown in Fig. 4.4. Here buffering is done at the input ports, and three input ports are shown with cells destined for output port 4. Only one of the three input buffers with cells for output port 4 will be able to transmit. The other input ports will send no cells, even though other cells in their queue are destined for other, perhaps idle, ports. Efficient use of memory buffers, e.g., via sharing across all output ports, is another consideration in buffer design in switches.

■ *Scalability of the architecture,* i.e., its ability to function at higher rates. Depending on the type of fabric, limits occur as a result of various fundamental technology characteristics, including memory and bus speed and physical layout restrictions such as the number of crosspoints on a circuit board.

- *Ability to efficiently support multicast and broadcast traffic.*
- *Ability to support various traffic classes,* such as those discussed in Sec. 2.3.5.

4.4.1 Shared Memory Approaches

The shared memory configuration is conceptually simple and powerful [Garc]. Shared memory approaches are similar in principle to a traditional time slot interchange (TSI) device typically used as part of a circuit switching system. As shown in Fig. 4.5, the TSI device reads a 24-octet frame (part of a DS1 rate 1.536-Mbps signal) into memory at a rate of one frame per 125 µs. After the entire frame is read into memory, the octets are read out in a new frame in a sequence defined by a control module. The 24 DS0 64-kbps input channels of the DS1 signal are thus interchanged in a manner defined by the control module. Figure 4.6 illustrates the process in more detail, with odd and even frames being subsequently read into one portion of memory and out the other. Items to note about this device are the following:

- The overall rate of input to the memory, 1.536 Mbps, is equal to the sum of the individual input DS0 channels. This required increase in memory speed above that of the input signals will sometimes be a limiting factor for shared memory ATM switches.

- The memory is dual ported, i.e., as one frame is being read in, the previous frame is being read out.

- A constant delay is introduced equal to the frame time (125 µs) of the DS1 signal.

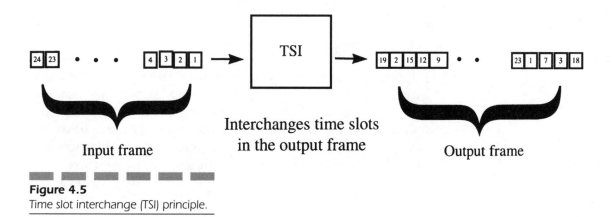

Figure 4.5
Time slot interchange (TSI) principle.

TSI

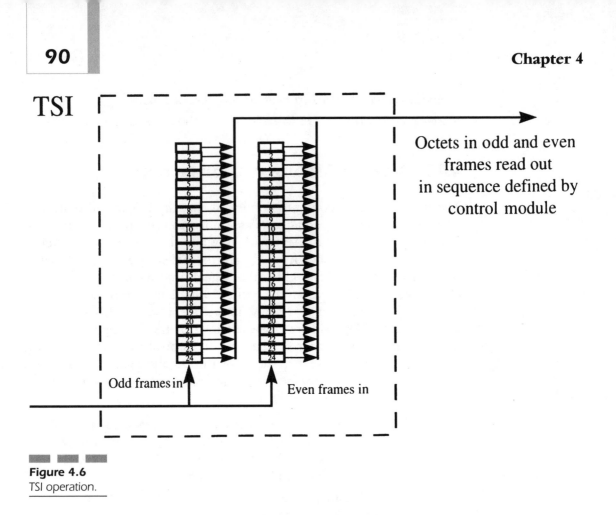

Octets in odd and even
frames read out
in sequence defined by
control module

Odd frames in

Even frames in

Figure 4.6
TSI operation.

An ATM shared memory switch is similar to the above TSI device, with the major exception that no contention occurs in the above example, since each of the input DS0 channels is mapped to a unique output channel. In the ATM switch case, as mentioned, contention occurs when two or more input ports attempt to route to the same output port.

The basic ATM shared memory switch architecture is shown in Fig. 4.7. Here input cells are converted from serial to parallel for input to the memory, time-division multiplexed to a higher data rate, and written sequentially into a dual-port memory. The memory essentially provides a form of output buffering for the switch. The output stream is formed by sequentially retrieving cells from the output buffers, one per buffer. A memory controller decides the order in which cells are read out of the memory, based on the internal routing tags assigned by the input functions.

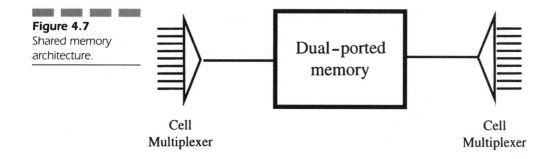

Figure 4.7
Shared memory architecture.

Cell
Multiplexer

Cell
Multiplexer

A design constraint of this approach is that the required memory bandwidth is twice the aggregate bandwidth of the input lines. For example, for a switching fabric with 16 input and 16 output ports, each operating at a line speed of 155 Mbps, a memory bandwidth of at least 4.8 Gbps is required. Memory systems used for a shared memory switching element are typically of the dual-ported type, since this type allows writing and reading to be performed independently and thus can achieve higher bandwidth. In addition, the higher memory bandwidth required for this configuration is generally obtained through the use of memory blocks at least 8 bits wide, and in many cases much wider.

Blocking Characteristics   Notice that this architecture is nonblocking; when cells from each input port are destined for a unique output port, it essentially acts in a manner similar to the TSI device discussed above.

Buffering Characteristics   The shared memory characteristics of this approach are particularly attractive because they can provide a *shared* buffer pool for all of the output ports, with the individual port buffers constructed via linked lists in the memory. All locations storing cells waiting in the queue for a given port are kept on a list. Cells in this linked-list, first-in, first-out queue are then read out and routed to the appropriate output port. This sharing of memory among the output ports can provide a significant decrease in the probability of dropping a cell as a result of buffer overflow. Figure 4.8 depicts two cases:

1. Complete partitioning of the memory to output ports (i.e., assigning a *dedicated* amount of the buffer memory to each of the output ports)

2. Full sharing of the memory among the output ports, as is obtained with the shared memory approach

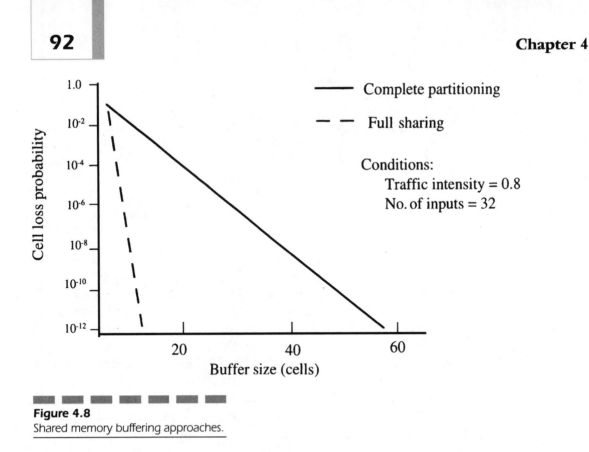

Figure 4.8
Shared memory buffering approaches.

The figure shows the dramatic reduction in buffer size needed to maintain a particular cell loss probability (CLP) for a particular traffic level with full memory sharing. For example, to attain a cell loss probability of less than 10^{-12} for the switch characteristics and traffic levels shown in the figure would require a 10-cell buffer size for full memory sharing and on the order of a 60-cell buffer size for complete partitioning.

A physical design advantage of this architecture is that buffer memory is associated with the switch fabric and thus is not needed on the physical input or output line cards. Line cards can therefore accommodate more ports in this design, reducing the number of physical input/output boards required for a switch of a certain size.

Scalability The scalability of this architecture is restricted by the fact that buffer memory must operate faster than the port speed (as in the TSI case). Since memory speed is physically limited, this places a corresponding limit on the scalability of a single switch. Future switches will have the advantage, however, of riding the increasing speed of semiconductor memory.

To obtain more switch capacity, the interconnection of multiple switches is required. The feasibility of such switches in the range of 160 Gbps has been reported in the industry [Tele].

Multicast and Broadcast   Multicast and broadcast cells can be handled via separate queues in the shared memory. Routing information is used to deliver multiple copies of the same cell to the specified output ports, so that only a single cell must be stored in buffer memory. By contrast, some other techniques require storage of multiple copies of multicast and broadcast cells at the output buffers.

Handling Various Traffic Classes   The basic linked-list buffer design can be further enhanced to provide for the various service classes. Separate buffers (still fully shared) can be allocated for the various service classes and a priority queuing discipline applied, with constant bit rate (CBR), real-time variable bit rate (rt-VBR), non-real-time variable bit rate (nrt-VBR), unspecified bit rate (UBR), and available bit rate (ABR) traffic assigned different priorities.

Sample Industry Design   A sample industry design employing this architecture is the Cisco Systems LightStream 1010, shown in Fig. 4.9. Cisco's LightStream 1010 switch fabric is based on the shared memory architecture with a total buffer capacity of 64K cells. The shared memory also supports programmable queue lengths for four classes.

This switch supplies 32 full-duplex ATM ports, each operating at the STS-3c rate of 155.52 Mbps, for a total throughput of about 5 Gbps. Four ports can be combined to provide a single STS-12c rate interface at 622 Mbps.

4.4.2   Shared Media Architectures

Like the shared memory approach, the shared media approach multiplexes the incoming cells into a higher-speed stream and then demultiplexes the stream into single streams for each output line. However, instead of storing the cells in a common memory, this approach utilizes a high-speed medium [usually a time-division multiplexed (TDM) bus or set of parallel buses] in which cells are multiplexed at a rate N times the rate of a single input line (N being the number of inputs).

Figure 4.10 illustrates the approach. Cells arriving at the inputs are sequentially broadcast on the TDM bus in a round-robin manner. At each output, address filters pass the appropriate cells to the output buffers, based on their routing tag. Demultiplexing of the cells is done at the output interfaces by

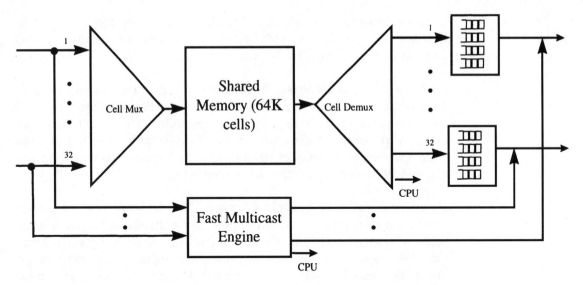

- Separate per-port output queues for CBR, RT-VBR, NRT-VBR and UBR/ABR
- Fast multicast engine replicates only pointers; physically stores only one copy of cell
- Switch processor also receives point-to-multipoint cells, allowing easier implementation of LANE

Figure 4.9
Cisco LightStream 1010 switch fabric.

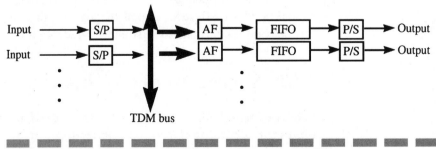

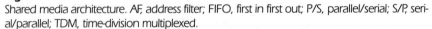

Figure 4.10
Shared media architecture. AF, address filter; FIFO, first in first out; P/S, parallel/serial; S/P, serial/parallel; TDM, time-division multiplexed.

the address filters. The filters read all the cells transmitted through the TDM bus, then transfer to the FIFO queue only those cells whose address matches that of the output port. Note that to eliminate input queuing with this design and eliminate the head of line phenomenon, the total bus bandwidth for a shared media ATM switch with N inputs will be N times the rate of a single input. This can be easily seen by considering the worst case of cells arriving at all N inputs being destined for a single output port. Here, to avoid input queuing, the shared media will need sufficient bandwidth to transfer all N cells to the output port in one switch cycle.

An important design consideration for this type of switch is the implementation of the high-speed bus and the bandwidth of the output memory and address filters (similar to the memory bandwidth design issue for the shared memory approach). The high-speed bus and the address filters have to operate at N times the input line speed. For example, in a switch with 16 inputs each at 155 Mbps, the high-speed bus and filters need to operate at 2.4 Gbps or greater. The memory bandwidth would need to operate at somewhat more than this rate to accommodate both reading and writing of cells. Again, that these rates are required can be seen by considering the case in which all cells are destined for a single output. The address filter would pass all cells in this case, and the memory bandwidth would need to be high enough to handle all incoming cells and deliver cells to its output port.

Many smaller carrier-type switches, in particular edge switches in the 5 to 10 Gbps range, use a single passive bus approach. Usually, the backplane serves as the bus and is equipped with the appropriate connections into which individual line modules are plugged, each module comprising an input and output port.

Blocking Characteristics   Similar to the shared memory approach, this technique is nonblocking.

Buffering Characteristics   Unlike in the shared memory approach, output buffers are not shared, and thus more buffers will be required for the same cell loss rate.

Scalability   Again similar to the shared memory approach, this technique is restricted by the fact that the shared media and buffer memory must operate faster than the input port speeds. One solution to this problem is to implement the cell switching fabric based on a bit-sliced scheme, in which the incoming bit streams are split into a number of parallel streams, each one connected to a different parallel subswitch [Toba]. Each parallel subswitch consists of its own shared bus, address filters, and FIFO queues running at reduced rates and thus can constitute a basis for implementing larger switches.

Multicast and Broadcast   The broadcast-and-select nature of this approach makes support of multicasting and broadcasting features straightforward.

Handling Various Traffic Classes   As in the shared memory approach, priority functions in the shared media approach are provided by organizing the output queues for the individual output lines into several priority classes. A priority field in the internal tag supplies the information concerning which specific priority class the cell belongs to and thus in which queue to store the cell.

Sample Industry Design   The Fore Systems ASX 200 family of ATM switches includes a range of switch products designed for both the carrier edge switch market segment and the enterprise backbone segment. All of Fore's ASX-200 switches are based on a nonblocking, time-division multiplexing architecture. Per-port transmission speeds up to 622 Mbps are supported. Interfaces ranging from T1 rates to OC-12c (1.544 Mbps to 622 Mbps) are available on a variety of media.

A typical configuration is shown in Fig. 4.11. Note the following about this configuration:

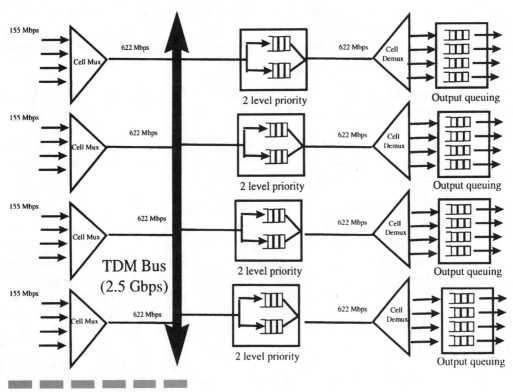

Figure 4.11
Fore Systems ASX-200 switch fabric.

■ The aggregate throughput is 2.5 Gbps, the speed of the bus.

■ Two stages of output buffering are employed. One stage provides a simple two-level priority queue, suitable, for example, for the support of two classes of service. The second stage provides a further level of queuing on a per virtual circuit basis.

Fore Systems also provides an ASX-1000 switching system with the following improvements over the ASX-200: Four 2.5-Gbps parallel buses are employed, raising the aggregate throughput to 10 Gbps, and the first stage of buffering provides a four-level priority queue rather than the two priorities of the ASX-200.

4.4.3   Space-Division Architectures

This type of architecture does not multiplex the incoming traffic into a single stream, as occurs in the shared memory and shared media (bus) architectures. Rather, a space-division switch has multiple concurrent paths, with the same data rates, to route cells from fabric inputs to outputs. Thus, with this approach it is no longer necessary for the switch to have a memory running at multiples of the line speed. However, this architecture has the major disadvantage of internal blocking, which can affect the throughput of the switch. Here, internal blocking occurs when two or more cells need to access the same internal link within the fabric. As a result, buffering has to be implemented at the location where the blocking occurs or upstream of the location, e.g., at the input of the switch, and not at the output as with the shared memory and shared media approaches.

Common space division architectures are those associated with crossbar fabrics and those constructed of multistage interconnection networks (MINs). Examples of these classes are discussed in the following sections.

Crossbar Fabrics   A crossbar fabric consists of an array of N^2 crosspoint switching elements, that is, one switching element for each input-output pair, as depicted in Fig. 4.12. This fabric can be implemented by using two states of the switching element, cross state or bar state, as shown in the figure. With N^2 switching elements, this technique allows for a nonblocking architecture; to switch cells from input port i to output port j simply requires closing the i,j switching element.

Items to note related to the use of this architecture for cell switching functions include the following:

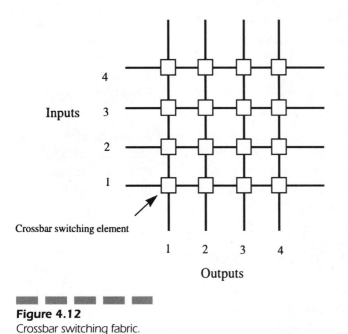

Inputs

4

3

2

1

Crossbar switching element

Outputs

Figure 4.12
Crossbar switching fabric.

- N² switching elements
- Switching element states:

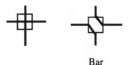

Bar

- The speed of setting the switching elements is one limiting factor in the throughput of the switching fabric.

- The mechanism for controlling the setting of the switching elements is also a key design element. In fact, to reduce the complexity of controlling the switching elements, a self-routing property can be employed. Here, as a cell is propagated through the crosspoints, an internal routing tag, assigned for example by the input functions, can serve to steer the cell properly through the switching array. As an example, Fig. 4.13 shows a 3×3 switching array and the paths taken to route cells through the array. Note that in the figure, no two cells were destined for the same output; if they were, cell buffering or dropping would occur.

- As mentioned above, buffering would be required in a crossbar fabric. There are two possible locations for the buffers.

  1. *At the input ports.* When a cell reaches a switching element and is blocked (e.g., by a previously arriving cell), a signal is returned to the input and the cell is placed in the input queue. A number of techniques for this signaling are possible. Since this approach is based on input buffering, it suffers from the head of line blocking phenomenon, discussed previously.

2. *At all crosspoints.* A crossbar switching fabric with buffer memory at the crosspoints eliminates the previously mentioned problems with input buffering. In effect, this approach is similar to one that uses output buffering except that the buffers are distributed along the crosspoints at the output line rather than at the output line itself. For example, in Fig. 4.13, the queue for a particular output is distributed over the three crosspoints located directly above that output. An advantage of this approach is that the required memory bandwidth at the crosspoints is only twice the bandwidth of each input port. A drawback of the approach is that there is minimal memory sharing, with the separate buffers at the crosspoints all being independent. In addition, combining the crossbar arrangement with the required address filters, memory, and gating functions requires a significant increase in integrated circuitry real estate for a practical implementation.

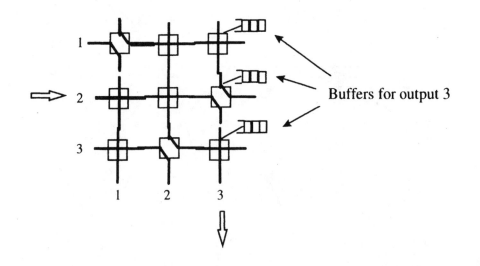

- Buffers for outputs are distributed over vertical crosspoints; only buffers for output 3 are shown
- Limited buffer sharing

Figure 4.13
Sample cell flow (3×3) crossbar fabric.

Multistage Interconnection Networks The switching fabrics described above can be interconnected to construct larger switching fabrics. Generally, most ATM switching systems that handle more than a total aggregate throughput of 5 Gbps require a switching fabric consisting of multiple interconnected switching fabrics.

The MINs are generally constructed using basic building blocks of 2×2 crosspoint switching elements, as depicted in Fig. 4.14. The individ-

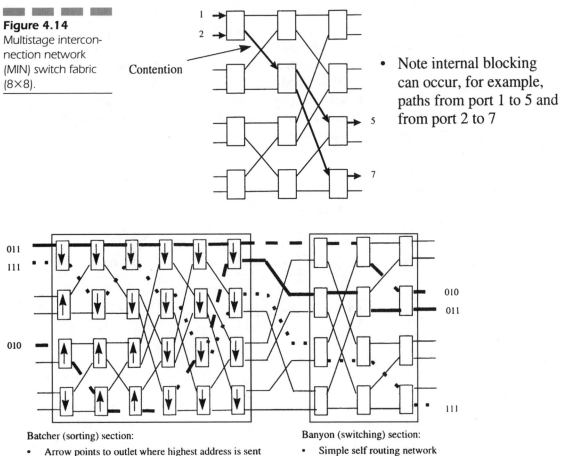

Figure 4.14
Multistage intercon-
nection network
(MIN) switch fabric
(8×8).

Contention

• Note internal blocking can occur, for example, paths from port 1 to 5 and from port 2 to 7

Batcher (sorting) section:

• Arrow points to outlet where highest address is sent

• If only a single cell is input to a sorting element, it is considered the lowest value

• If cells at 2 inputs are destined for the same output, duplicates are fed back to later slots

Banyon (switching) section:

• Simple self routing network

• No Head of Line collision (Batcher section eliminates duplicate output cells

Figure 4.15
Batcher-Banyan switch fabric.

ual elements are able to take two states, cross state and bar state, like the elements of the crossbar switches discussed in the previous subsection.

Comparing this fabric with the crossbar fabric, fewer switching elements are needed to implement the MIN design. The regular and modular structure of this network, which allows the implementation of large networks from smaller ones, has made it most suitable for implementing space-division switches. While, in general, MINs have the drawback of internal blocking, certain types of MINs, e.g., Banyan-network-based architectures, have been developed to solve this drawback and the resulting throughput limitations. One of these solutions is the Batcher-Banyan switching fabric, presented in the following subsection.

Batcher Banyan   One special case of a space-division fabric is the Batcher-Banyan fabric. It belongs to a special class of switching fabrics that do not have internal contention of cells passing through the switch. Thus, since no contention exists in the switch, there is no need to provide internal buffering. Figure 4.15 indicates the structure of the Batcher-Banyan fabric. The fabric consists of two sections.

1. The Batcher section sorts the cells according to their destination address. The result of this sorting operation is that the cells with the lower address are sent to the upper outlet of the Batcher section. At the output of the Batcher section, cells are sorted for input to the Banyan section.

2. The Banyan section is a simple self-routing network that guarantees that all cells will arrive at the destination indicated by the routing information in the header. There is no head of line collision at the entrance of the Banyan section, and since the Banyan section has no internal blocking, there will be no further blocking or output contention.

Numerous scientific papers on Batcher-Banyan fabrics have been published; however, actual switches intended for the market rarely employ this technique.

4.4.4   Summary

Table 4.1 provides a summary of the major characteristics of ATM cell switching fabrics discussed above. Table 4.2 lists a sample of the types of ATM cell switching fabrics used in industry products.

TABLE 4.1

Cell Switch Fabric
Characteristics

Switching Architecture	Key Design Issues/Disadvantages	Solutions/Advantages
Shared memory	Cell processing time, memory bandwidth, memory access time, memory size	Complete queue partitioning
Shared medium	Similar to shared memory, plus implementation of high-speed bus	Bit-sliced organization
Space division 1. Crossbar 2. MIN	Internal blocking	Buffering at the location of the contention or upstream

TABLE 4.2 Types of Cell Switching Fabrics in Use in the Industry

Shared Memory	Shared Media	Shared Media (Parallel Bus)	Crossbar	Banyan or MIN
Alcatel ISE	ADC Telecomm	Cascade 500	3Com CellPlex-7000	Alcatel Data Net
Bay/Centillion	Cisco LightStream 100	DSC MegaHub	ADC Fibermux	Bay Lattis Cell
Cisco LightStream 1010	Fore ASX-200	Fore ASX-1000	BBN Gigabit Router	IgT
CNET Prelude	NEC Models 5, 10, 20	GTE SPANet	Cisco BFR/GSR	
Hitachi	NET/Adaptive ATMX	Newbridge 36170	DEC Giga Switch/ATM	
IBM	NTI Passport	NTI Gateway	Fujitsu Fetex-150	
Lucent Globeview	Cisco/Stratacom IGX	TRW BAS-2010c	GDC Apex	
NTI Concord			Newbridge 36150	
Siemens			Scorpio/U.S. Robotics	
Toshiba			UB GeoSwitch-150	

4.5 Connection Management

4.5.1 Background

Connection management functions are concerned with the procedures for establishing and maintaining connections through the ATM switching system. As a result of establishing these connections, the connection management function establishes and maintains a connection table that reflects parameters associated with the connections. An example of entries in such a table is shown in Fig. 4.16. This table is accessed as ATM cells are switched on established connections. The connection management function interacts with the CAC portion of the traffic management and congestion control function, discussed in Sec. 4.6, in determining whether or not a connection request should be honored. Options for the physical placement of the connection management function vary. As discussed in Sec. 4.2, lower-level connection management functions such as the handling of signaling cells can be distributed to individual input modules or to groups of input modules.

As part of the connection management function, signaling information is exchanged at various interfaces in the ATM network. These interfaces include the following, which are shown in Fig. 3.1.

■ Between an ATM switch and an end user over the User-Network Interface (UNI)

■ Among ATM switches, typically over a Broadband Intercarrier Interface (B-ICI) in a public carrier network or between carrier

Figure 4.16
Connection table in
an ATM switch.

	Input		*Connection*		*Output*		
Port	VPI	VCI	Rate	Class	Port	VCI	VPI
1	0	50	100 kbps	CBR	17	0	27
1	0	62	8 Mbps	VBR	5	0	22
1	0	62	8 Mbps	VBR	6	1	384
2	2	17	50 Mbps	CBR	1	0	20

- Above shows a simplified connection table in an ATM switch

- First entry describes a 100-kbps constant bit rate (CBR) connection received on port 1 using VPI=0 and VCI=50, transmitted on port 17 using VPI=0 and VCI=27

- Second and third entries describe a point-to-multipoint variable bit rate (VBR) connection

- Fourth entry is another CBR connection

networks, and over the Private Network-Node Interface (P-NNI) in private networks

The complexity of implementing connection management functions in an ATM switch is due in part to the need to support these functions over all of these interfaces and to provide interworking between the signaling procedures involved on the various interfaces.

The remaining subsections discuss these differing signaling interfaces. While they are similar in nature, there are differences among them; for example, signaling procedures at the UNI are not symmetrical, since this interface involves a user's equipment interfacing to a network, whereas the B-ICI interface is of a peer-to-peer nature, involving the interface of two networks.

4.5.2   User-Network Interface Signaling

Signaling for call control at the UNI evolved from signaling procedures for controlling narrowband channels via ITU-T Recommendation Q931. Call control at the ATM UNI added signaling for the control of the variable-bandwidth channels associated with ATM; this is published in ITU-T Recommendation Q2931. This recommendation specifies the message sets used in controlling call setup and teardown and describes the parameters included in the signaling messages and their codings. Examples of parameters are the traffic descriptors for the requested connection. The various call states and their progression are also described. As previously mentioned, the signaling channel on this interface is identified by VPI=0, VCI=5. The ATM Forum UNI signaling specification [af-sig-0061.000] is based on the Q2931 specification and addresses both point-to-point and point-to-multipoint connections. The specification applies to the UNI in both public and private networks.

4.5.3   Broadband Intercarrier Interface Signaling

Call control signaling over the B-ICI [af-bici-0013.003] differs from that over the UNI primarily in the following areas.

- As mentioned, the interface is symmetric and peer-to-peer in nature.
- Because of its use in public networks, some additional reliability features are provided.

■ It is architected to evolve to the use of the carriers' imbedded signaling network (Common Channel Signaling System 7, or SS7). This network is a separate packet-based network that carries signaling information among the carriers' switches.

Protocols defined for this signaling interface are shown in Fig. 4.17. This protocol stack is based on a subset of the SS7 family of protocols and services for the signaling network. The overall objective of the subset of SS7 protocols in Fig. 4.17 is to provide reliable sequenced transfer of signaling messages without necessitating use of the SS7 network.

Again, signaling messages are exchanged directly among the ATM switches using a signaling ATM virtual channel. The Signaling AAL (SAAL) has the same structure as the UNI, i.e., it is based on AAL 5. The Message Transfer Part (MTP) protocol is an SS7 protocol that minimizes message loss and missequencing. A simplified subset of MTP is specified for initial use on the B-ICI, essentially for switch-to-switch signaling channel connections using multilink point-to-point diverse physical facilities and not requiring use of an SS7 network. The connection control messages and procedures are based on the B-ISDN User Part (BISUP), specified in ITU-T Recommendation [Q2764].

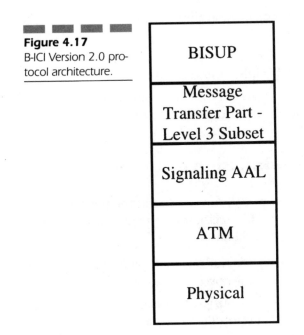

Figure 4.17
B-ICI Version 2.0 protocol architecture.

4.5.4   Private Network-Node Interface

Signaling for call control between private networks or between ATM switches in private networks, as specified by the ATM Forum [af-pnni-0055.000], is based on a subset of UNI signaling with changes to make the procedures symmetric. P-NNI signaling also makes use of route calculations maintained by the P-NNI routing functions.

4.6   Traffic Management and Congestion Control

The essential purpose of the traffic management and congestion control function is to assert controls to avoid the onset of congestion in an ATM switch (generally caused by high buffer occupancy) and, if congestion is encountered, to take actions to alleviate the condition (e.g., drop all incoming cells).

A number of traffic management and congestion control techniques have been defined for ATM switches and are discussed in this section. These actions and controls take place on a variety of time scales.

- Those associated with cell transmission times (on the order of microseconds)
- Those associated with cell round-trip times (on the order of microseconds for LANs and milliseconds for WANs)
- Those associated with the interarrival times for connection establishment requests (on the order of milliseconds)

4.6.1   Connection Admission Control

CAC is the set of actions taken to determine whether or not a connection request should be accepted. This decision is generally based on two somewhat complementary sets of criteria:

1. Whether the required resources are available to provide the end-to-end Quality of Service (QoS) objectives for the connection
2. Policy-related matters, such as user-configured access lists based on source or destination addresses, protocol types, time of day, etc.

A connection is established if the network has enough resources to provide the QoS requirements of the connection request without affecting the QoS provided to connections already established in the network. Resource sufficiency is based on the requested ATM service category, the identified traffic descriptors, the specified QoS parameter values, the resources currently allocated to existing connections, and the current load on the network.

The admission decision process is quite complicated, since a CAC procedure needs to make accurate decisions while balancing somewhat conflicting factors, viz., the service requirements of all connections need to be met, but high utilization of network resources needs to be achieved. In addition, any algorithm used for this purpose needs to operate in real time. To speed this decision process, a number of algorithms are used to calculate what is usually called the "equivalent bandwidth" needed by all traffic sources. These algorithms provide probabilistic estimates of the bandwidth required to meet the QoS of all traffic sources, including the requested connection. These algorithms are chosen to be computationally efficient and easily implementable and are the focus of much current research [Gelenbe].

The CAC interacts with the routing function to select a route, allocates bandwidth, and assesses the relative priority of the new and existing traffic. These actions are performed directly by the ATM switch when a switched virtual connection (SVC) is requested and by network management procedures when a permanent virtual connection (PVC) is being established. Based on the specific CAC algorithms in place, a connection request will progress through the network only when sufficient resources exist at each successive switch in the end-to-end connection.

As part of the connection admission process, the CAC also negotiates a "traffic contract" with users requesting new connections. The traffic contract is an agreement between the user and the network that describes the negotiated characteristics of the connection and is the key driver for other traffic management and congestion control functions discussed in this section.

There are a number of physical design issues related to the placement and distribution of the CAC functionality. If the CAC is centralized, a single processing unit would perform admission decisions and resource allocation decisions for all connections in the switch. For large switch sizes, this could result in a processing bottleneck and may limit the connection rate of the switch. CAC functions may also be distributed to blocks of input modules, in which case a particular CAC processor would have responsibility for connections involving only a set of input

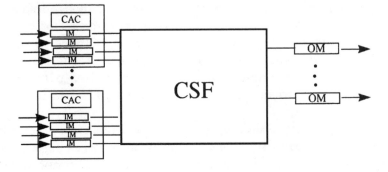

- Centralized connection admission control (CAC) is simple to implement but could be bottleneck for large switches since it handles CAC for all input modules.

- Another option is to distribute the CAC functions to groups of input modules; more complex since state information needs to be updated, distributed and coordinated among the CACs.

Figure 4.18
Distribution of connection admission control (CAC). CSF, cell switch fabric; OM, output module; IM, input module.

ports. This alternative, shown in Fig. 4.18, is harder to implement, since information must be communicated and coordinated among the various CAC modules.

4.6.2 Usage Parameter Control

Usage parameter control (UPC) is defined as the set of actions taken by an ATM switch to monitor and control traffic at the UNI on each active connection. The objective of UPC is to protect the network and its users from the malicious or unintentional misbehavior of another network user, for example, as a result of malfunctioning hardware or software.

The UPC function checks the compliance between the monitored or measured parameter values and the negotiated parameter values (as reflected in the traffic contract), then takes appropriate actions in the case of violation. Cell conformance is generally checked via one or more applications of a generic cell rate algorithm (GCRA), commonly referred to as the leaky bucket algorithm. One of the basic functions of the

GCRA is to allow a certain amount of cell "clumping" associated with the normal variation in cell interarrival times. This variation is due to all of the factors described in earlier sections (e.g., cell assembly, cell multiplexing, addition of OAM cells) and creates a temporary situation in which the peak cell rate is exceeded for a period of time (of course, there will be times when the rate falls below the peak cell rate). If this clumping persists beyond that allowed, actions such as "tagging" the cell by setting the CLP bit in the header or discarding the cell can be taken. For the ABR service category, the traffic source is expected to adjust its cell emission rate based on feedback from the destination and/or intervening switches. The cell conformance definition for ABR requires compliance with the source, destination, and switch behaviors specified by the ATM Forum [ATMF-TM].

4.6.3   Cell Loss Priority Selective Discard

One bit is defined in the ATM cell header for the explicit indication of cell loss priority (CLP). The CLP bit can be used to generate different priority cell flows within a connection. The users, applications, end systems, or an enforcement action of UPC upon nonconforming cells (e.g., tagging as discussed above) may set CLP=1, making these cells more likely to experience loss during network congestion. A selective cell discarding scheme based on CLP can be applied to deal with network congestion at the point of congestion—usually an ATM switch queuing point. For example, when the switch port queue occupancy reaches a management-defined threshold level, only CLP=0 cells are allowed to enter the system and CLP=1 cells are discarded.

4.6.4   Selective Packet Discard

While short, fixed-length cells facilitate high-speed hardware switching, most if not all data units from the higher layers are variable-length packets. Individual cells are placed into packets via the segmentation and reassembly (SAR) process discussed in Chap. 2. For most applications, such as those on LANs, when one or more cells are dropped by the network, the corresponding corrupted packets become useless and will be discarded by the higher layers. To maximize the number of completely delivered packets, cell discard schemes are defined to intelligently and selectively discard cells belonging to the same packets. Two of these schemes are the following:

Tail packet discard.   Whenever one cell has been discarded, tail packet discard discards all subsequent cells of the packet except the last one, indicated by the service data unit type bit in the payload type (PT) field of the cell headers discussed in Chap. 2. This last cell acts as an indication to the higher layers in the end system to reassemble the higher-layer protocol data unit (which will subsequently, of course, be discarded).

Early packet discard.   Early packet discard (EPD) is a technique in which the switch starts to discard all cells, except the end-of-packet cells, from newly arriving packets when the switch buffer queues reach a threshold level. With this technique, if cells from a packet have already entered the buffer, all remaining cells of the packet are allowed to enter (if sufficient buffer space is available). Note that whereas tail packet discard will reduce the effect of cell discard on corrupted packets, early packet discard will eliminate it, since effectively all cells from a packet (except the last) will be discarded.

The goal of these packet discard schemes is to prevent frequent buffer overflows by dumping corrupted or complete packets from the rapidly filling buffers rather than dumping random cells, which could affect many packets. When a small number of packets are dropped instead of cells from a large number of packets, occasional buffer overflows do not have serious negative effects on the end-to-end throughput.

4.6.5   Congestion Notification and Flow Control

Within the network, a network element in a congested state can set the standardized explicit forward congestion indication (EFCI) code point in the PT field of a cell header as an indication of congestion so that it can be examined by the destination end system. The end system may optionally react to the EFCI as part of an end user's feedback control mechanism (such as TCP window flow control) to minimize the impact of network congestion.

The ABR service specifies a set of mechanisms and a process that both switches and ATM end systems can use to *dynamically* regulate the amount of traffic sent on ABR connections. Source end systems periodically use resource management (RM) cells, indicated by another code point in the PT field, to probe the network state for information on bandwidth availability, state of congestion, and impending congestion.

RM cells are intermixed with user data cells along all ABR connections to estimate bottleneck service rates, including feedback delays. RM cells are received by the destination end systems and returned to the source on the same connection, for a closed loop that indicates whether intermediate switches have experienced congestion. Examples of the use of RM cells include the following:

■ The switch can set the EFCI state in the headers of forwarded cells to indicate congestion. The destination receiving an EFCI flag sets the congestion indication (CI) bit in the backward RM cells to indicate congestion to the source.

■ The switch can set the CI bit in both forward and backward RM cells. This method, because of its ability to use the backward RM cells to send the congestion indication (a technique also known as backward error congestion notification) can greatly reduce feedback delays and deliver better performance.

■ The switch can reduce the explicit rate field of the forward or backward RM cells to explicitly indicate the rate at which the source may send. While this scheme may provide better performance, it requires significant complexities on the switch to implement.

In each case, the source system increases or decreases its rate depending on the type of feedback received.

4.7   Routing Functions

4.7.1   Overview

Routing functions are concerned with determining paths through the network and maintaining routing tables in the network nodes that reflect these paths. Routing functions can best be thought of as consisting of two separate processes.

1. The collection of data related to the topology of the network and the attributes of the network links (utilization, cost, bandwidth, etc.). Note that this first process can occur only occasionally, e.g., performed manually by a network administrator, or on a continuous ongoing basis via information exchanges among the network nodes.

2. Use of the above data to calculate network paths, via routing algorithms, and populate node routing tables. Here again, this could be an occasional or a dynamic, adaptive process.

Routing strategies differ primarily based on the time scales associated with the above processes. This breakdown is shown in Fig. 4.19, which lists the major characteristics of the fixed and adaptive strategies. In general, the tradeoff with routing strategies is the increased network resource utilization obtained at the cost of added complexity of the routing strategy. In practice, it is not uncommon to attain 20 to 30 percent improvement in resource utilization with more complex strategies. Also shown in the figure is a routing strategy based on a "flooding" technique used in some packet networks. In this case, the routing strategy is simply for the nodes to flood packets to all outgoing trunks, resulting in a very robust network, but one in which resources are not highly utilized.

A number of other characterizations of the routing function are applicable [Stallings]. These include

- The location for the collection of the network data and calculation of routing paths. This can range from occurring at one central location to being distributed among all network nodes.

- The source of the network information. This can range from no information being collected, as in the case of flooding, to information being collected only from adjacent nodes, to information being collected from all nodes.

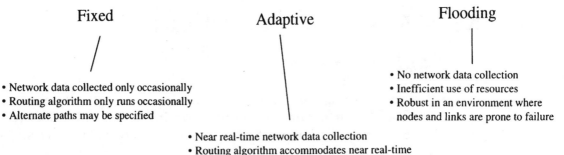

Fixed
- Network data collected only occasionally
- Routing algorithm only runs occasionally
- Alternate paths may be specified

Adaptive
- Near real-time network data collection
- Routing algorithm accommodates near real-time changes in the network (e.g., failures, congestion)
- Increased processing burden on nodes (if distributed)
- Increased traffic (related to network information)
- Increased network resource utilization

Flooding
- No network data collection
- Inefficient use of resources
- Robust in an environment where nodes and links are prone to failure

Figure 4.19
Sample routing strategies.

4.7.2 Routing Strategy Examples

The Public Switched Telephone Network The public switched telephone network (PSTN) is a circuit-switched network that traditionally uses a simple fixed, static routing approach based on a tree structure. In the PSTN, the switching nodes construct a path starting at the calling party and tracing up the tree hierarchy (where the levels in the hierarchy are referred to as classes) to the first common node, then tracing down to the called party. Some improvements that have been introduced in this approach include the following:

- Additional, nonhierarchical direct routes have been added where sufficient traffic has warranted.

- Alternative routing schemes were introduced to allow a node the choice of a number of predefined routes. The choice of route can be made in a fixed sequence or can be based on factors such as time of day and historical traffic conditions.

- More recently, dynamic, adaptive routing schemes have been added to enable the switching nodes to react in near real-time to changing traffic patterns on the network. Here the switching nodes exchange information related to local traffic conditions among themselves.

Internet Routing To simplify the exchange of network information in an Internet environment, the nodes (routers) are administratively assigned to routing domains, and procedures for intradomain and interdomain routing functions are defined. Most nodes are concerned only with intradomain routing and defer interdomain routing functions to a small number of designated interdomain nodes. Examples of routing protocols for these two purposes are the Open Shortest Path First (OSPF) protocol, defined in RFC 1583, for intradomain routing and the Border Gateway Protocol (BGP) for interdomain routing.

Most routing algorithms used to calculate the paths in packet-switched networks are variations of one of two common algorithms, the Dijkstra algorithm [Dijk] and the Bellman-Ford algorithm [Bellman].

ATM Networks As outlined in Sec. 3.3.2, the ATM Forum has defined both a fixed routing strategy, defined as part of the Interim Interswitch Signaling Protocol (IISP), for near-term availability and an adaptive (and more complex) strategy, termed the Private Network-Node

Interface (P-NNI). The major functions of the P-NNI routing protocol are the discovery of neighbors, the discovery of the status of links connecting to neighbors, the exchange of topology state information among the ATM switch node neighbors, and the synchronization of the network topology databases in the ATM switch nodes. The specification also provides a sample algorithm for the calculation of the network paths.

4.8   Switch Management Functions

A convenient and exhaustive characterization of the operation and maintenance functions associated with telecommunications networks and equipment is provided by the ITU-T Recommendations [M.3010] [M.3400]. This characterization, termed the Telecommunications Management Network (TMN), is used extensively by carriers in planning and implementing the management aspects of their networks and is finding increased acceptance by large corporations as well. The TMN provides an architectural framework for the management and operation of telecommunications networks and services and introduces five functional operations-related areas.

■ Configuration management is concerned with entering and updating information in databases within the network elements (e.g., ATM switches).

■ Fault management functions include identifying, verifying, and isolating troubles.

■ Performance management is concerned with the evaluation of and reporting on the operation and status of the network elements and would include, for example, monitoring the performance characteristics of the ATM switch.

■ Accounting management functions are those associated with the collection of accounting records and would enable the use of the ATM switch to be measured and costs allocated for such use.

■ Security management provides for the prevention and detection of improper use of network resources and services.

Conceptually, these functions are performed via management entities in the management planes shown in Fig. 2.1. These management entities interact with layer entities such as the ATM Layer entity (performing functions associated with the ATM Layer services) and the SONET Layer entities (performing functions associated with the SONET Layers). This relationship between the user and control plane entities and the management entities is shown in Fig. 4.20 for the fault management function. Here, the layer management entities collect information on events indicative of potential equipment or network faults, perform analysis, and forward fault indications to the management plane. The management plane may perform a correlation of fault indications received from all of the Management Layer entities in the switch (SONET Layers, ATM Layer, AAL, etc.) and forward only the root cause information to the Network Management System (NMS).

Note that the TMN addresses issues beyond those associated with a single network element, such as an ATM switch. For example, it addresses issues associated with the management of a collection of

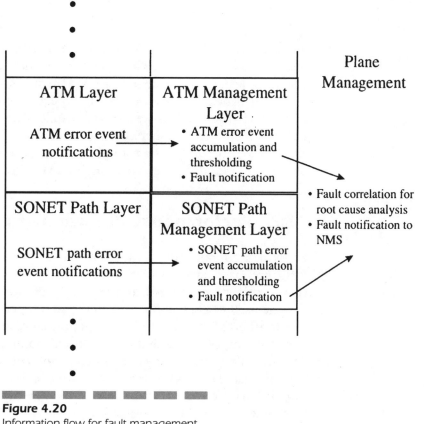

Figure 4.20
Information flow for fault management.

network elements via an NMS, the management of services provided by network elements, and even business and policy issues related to network management. In this section we present a brief description of the management functions associated with a single ATM switch following the TMN model. To aid in the distribution of management functionality between the ATM switch itself and its associated NMS, the ATM Forum has defined a standard interface between these components, as shown in Fig. 3.1. This interface definition consists of a standard vendor-independent view of the management information exchanged between the switch and the NMS and a protocol specification for the interaction. See Sec. 3.3.2 for the status of this work in the ATM Forum.

4.8.1 Configuration Management

Configuration management functions for an ATM switch include the ability to receive initial software downloads, activate the software, and maintain a record of the version and update level of the software. On an ongoing basis, the ATM switch needs to provide access to information relating to its service feature settings. ATM switches also have the ability to provide self-inventory features relating to equipment configurations and to report changes in the state of certain equipment types, e.g., a change in state from "enabled" to "disabled." Overall network planning and design is also considered part of the configuration management process, but this of necessity includes more than a single ATM switching system.

4.8.2 Performance Management

Performance monitoring is the major activity performed under the performance management function. Performance monitoring is the nonintrusive collection, analysis, and reporting of performance data. These data are used to assess and maintain the switch performance.

For an ATM switch, the major performance monitoring functions would be physically located on the input module, as shown in Fig. 4.2. This module would continuously detect and collect statistics on events occurring at the SONET level, such as out of frame, protection switching, or parity error events, and ATM-level events such as HEC errors. Generally the number of events is accumulated for a setable time interval, e.g., 15 min, and reported to the NMS.

4.8.3 Fault Management

The ATM switch provides basic input to this function via the detection of failure events and reporting of them to a NMS. To avoid spurious failure event indications, failure event reporting is generally based on a count of errors that reaches or exceeds a setable threshold. In addition, hardware and software checks are performed in a background mode on a scheduled basis. When such checks fail, the results are also reported to the NMS.

The ATM switch also must assist in the localization of network faults. Here again, upon request or on a scheduled basis, hardware and software diagnostics or audits run within the switch, with results reported to the NMS. For example, an NMS may request a switch to participate in fault localization via use of the OAM fault management cells discussed in Chap. 2.

As part of the fault management process, the switch may provide automatic restoration features. In addition to the SONET restoration procedures, which may be invoked at the SONET Line or Path layers, restoration may also be accomplished at the ATM Layer, e.g., via the alternative routing procedures discussed in Sec. 4.7.

4.8.4 Accounting Management

Basic usage information is collected as part of this function to provide support for activities such as billing for resource usage and planning/engineering for switch/network additions and upgrades. Physically, the collection of the raw data on cell usage is most logically associated with the switch input modules, where the cells are prepared for forwarding through the switch fabric. Longer-term storage and preliminary summarization of the data prior to shipment to automated systems involved in the billing or network engineering process may be provided centrally in the switch.

4.8.5 Security Management

Security management is concerned with the prevention of unauthorized intrusions to TMN elements. Examples of functionality required are the authentication and validation of a user's or network operator's requests for access to network capabilities. As mentioned in Sec. 3.3.2, specification of these functions for ATM network elements is being addressed in the ATM Forum Security Working Group.

4.9   Service-Specific Functions

Service-specific functions may optionally be performed by the ATM switching system. Performance of such functions is accomplished through the termination or manipulation of higher-layer protocols associated with these functions. Examples of service-specific functions include the Frame Relay Service-Specific Convergence Sublayer (FR-SSCS), part of the AAL, for mapping frame relay data frames into ATM cells. Such interworking between frame relay service and ATM requires attention not only to the end user data but also to the interworking of the management and control features of the two technologies. For example, traffic and congestion control features need to interwork. In particular, the EFCI code point in the ATM cell header relates to the forward explicit congestion notification (FECN) and backward explicit congestion notification (BECN) of the frame relay service layer, and the CLP bit in the ATM header relates to the frame relay discard eligibility (DE) bit. In addition, frame relay service layer functions such as the use of flags for frame delimiting and zero insertion for avoiding false flags need to be terminated at the FR-SSCS, since these are unnecessary when the user data are encapsulated in AAL5, where a length field is used instead of flag delimiters. Finally, a mapping is needed between the virtual circuit identified by the frame relay data link connection identifier (DLCI) and that identified by the ATM VPI/VCI. Additional details of frame relay and ATM interworking will be covered in detail in later chapters.

4.10   References

[af-bici-0013.003] *BISDN Inter-Carrier Interface (B-ICI) Specification*, Version 2.0 (Integrated), December 1995.

[af-pnni-0055.000] *Private Network-Network Specification Interface*, Version 1.0, March 1996.

[af-sig-0061.000] *ATM User-Network Interface (UNI) Signaling Specification*, Version 4.0, July 1996.

[ATMF-TM] *ATM Forum Traffic Management Specification,* Version 4.0, af-tm-0056.000, April 1996.

[Bell] Bellcore, *Broadband Switching System (BSS) Generic Requirements,* GR-1110-CORE, Revision 3, April 1996.

[Bellman] L. Ford and D. Fulkerson, *Flows in Networks,* Princeton University Press, Princeton, N.J., 1962.

[Chen] Thomas M. Chen and Steve S. Liu, "Management and Control Functions in ATM Switching Systems," *IEEE Network,* July/August 1994.

[Dijk] E. Dijkstra, "A Note on Two Problems in Connection with Graphs," *Numerical Mathematics,* October 1959.

[Garc] Joan Garcia-Haro and Andrzej Jajszczyk, "ATM Shared-Memory Switching Architectures," *IEEE Network,* July/August 1994.

[Gelenbe] E. Gelenbe, X. Mang, and R. Onvural, "Bandwidth Allocation and Call Admission Control in High-Speed Networks," *IEEE Communications Magazine,* May 1997.

[Hels] T. K. Helstern and M. Izzo, "Functional Architecture for a Next Generation Switching System," *INFOCOM '90,* pp. 790–795.

[M.3010] *Principles for a Telecommunications Management Network,* ITU-T Recommendation 3010, October 1992.

[M.3400] *TMN Management Functions,* ITU-T Recommendation 3400, May 1996.

[Q931] *Digital Subscriber Signalling System No. 1 (DSS1)—ISDN User Network Interface Layer 3 Specification for Basic Call Control,* ITU-T Recommendation Q931, March 1993.

[Q2764] *BISDN, BISDN User Part—Basic Call Procedures,* ITU-T Recommendation Q2764, October 1994.

[Q2931] *Broadband Integrated Services Digital Network (B-ISDN)—Digital Subscriber Signalling System No. 2 (DSS2)—User Network Interface (UNI) Layer 3 Specification for Basic Call/Connection Control,* ITU-T Recommendation Q2931, February 1995.

[Stallings] William Stallings, *Data and Computer Communications,* 5th ed., Prentice-Hall, 1997.

[Tele] Kai Y. Eng, "Future Switch: An Inside Look at the Design and Operations of a Terabit ATM System," *Telephony,* Oct. 7, 1996.

[Toba] Fouad A. Tobagi, "Fast Packet Switch Architectures for Broadband Integrated Services Digital Networks," *Proceedings of the IEEE,* 78 (1), 1990.

Case Study:
The NCX 1E6
Multiservice
ATM Switch from
ECI Telematics Inc.

5.1 Company and Products Overview

5.1.1 ECI Telecom

ECI Telecom is dedicated to providing communications equipment and solutions to network service providers (NSPs) and large end users around the world. ECI's vision is to provide high-performance, low-cost network solutions that enable telecommunications users to increase capacity, improve quality, and maximize overall communications infrastructure effectiveness.

ECI Telecom provides a range of solutions embracing users' transmission, access, and service network switching needs, with particular focus on networkwide management of voice, data, fax, and video capabilities.

This chapter is focused on describing Telematics' NCX 1E6 asynchronous transfer mode (ATM) product.

5.1.2 NCX 1E6 ATM Switch

Telematics' NCX 1E6 is a flexible ATM switching platform in the NCX product line. With Subscriber-Oriented Networking at its core, the NCX 1E6 WAN switch is well suited for many central office and customer premises ATM applications. Designed for both adaptation and native ATM service termination, the NCX 1E6 delivers high-performance switching for access and backbone links ranging from DS0 to 155 Mbps on a fully redundant Network and Equipment Building Systems (NEBS) and ETSI compliant architecture.

5.1.3 Open Management System

Each Telematics product line is managed through Open Management System (OMS), a network management system based on Simple Network Management Protocol (SNMP) and Hewlett-Packard's OpenView distributed management platform. OMS provides a set of applications

This chapter has been prepared and supplied by Roy D. Rosner and David Boymel of ECI Telematics. Telematics' products enable a spectrum of networking applications ranging from narrowband packet and frame relay access to wideband frame relay and broadband/multimedia ATM switching networks.

specifically designed for element and service management of Telematics networks, with particular strengths in the areas of configuration, change, performance, and fault management. Through OMS, network operators are given an intelligent, easy-to-use, graphical interface for service provisioning, monitoring, and problem diagnosis to enable management of individual service-level agreements (SLAs).

5.1.4   Network Components

The NCX Next Generation Networks product family includes a range of multiplexers and switches designed to meet access, edge, and backbone applications in multiservice networks.

Figure 5.1 shows the typical deployment of NCX 1E6s in an ATM network.

Telematics classifies network element functions into the following generic areas, depending on functionality:

- Backbone switch
- Edge switch
- Access multiplexer

Backbone Switch   The NCX 1E6 can be configured as a backbone switch, designed exclusively to provide ATM switching. By definition, backbone switches typically are optimized for trunk services; that is, no subscriber access facilities are provided.

Figure 5.1

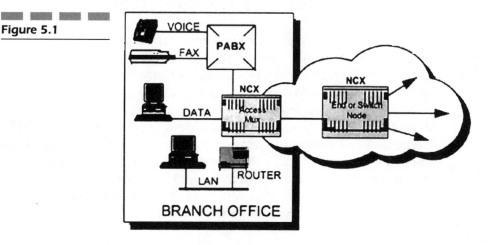

Edge Switch   The NCX 1E6 is strongly positioned as an ATM edge switch. Edge switches are functionally similar to backbone switches, but provide subscriber interfaces. Edge switches are designed with additional buffer capacity to provide a variety of service interfaces.

Access Multiplexers   Access multiplexers allow subscribers to access an ATM network through ATM and non-ATM interfaces. Access multiplexers typically are located at customer premises, and can be owned and operated by the customer or the NSP.

5.1.5   Connecting User Equipment

The NCX 1E6 supports the following subscriber interfaces:

- ATM User-Network Interface (UNI)
- Circuit emulation (voice, video, etc.)
- Bitsync (X.25, SDLC, HDLC, etc.)
- Frame relay
- Data support CBDS/SMDS
- PABX (G.704, G.732)

Figure 5.2 provides a view of the variety of applications that can be exploited through an ATM network by using the NCX 1E6 and other ECI Telecom network switching equipment.

5.1.6   Understanding the Subscriber

The driving force behind the design and marketing of the NCX 1E6 is to provide an ATM WAN switch that enables users to tailor individual, differentiated service contracts with network resource allocation that is closely matched to end-customer needs, traffic guarantees, and service pricing. The product design that enables such a proposition is very different from product designs oriented toward delivering only basic ATM services at a low equipment cost per port. WAN switches designed with equipment cost per port as the primary objective will not enable services to be provided or sold based on anything other than "best efforts" contracts. These best efforts products do not allow the full realization of ATM—the ability to deliver multiple narrowband, wideband, and

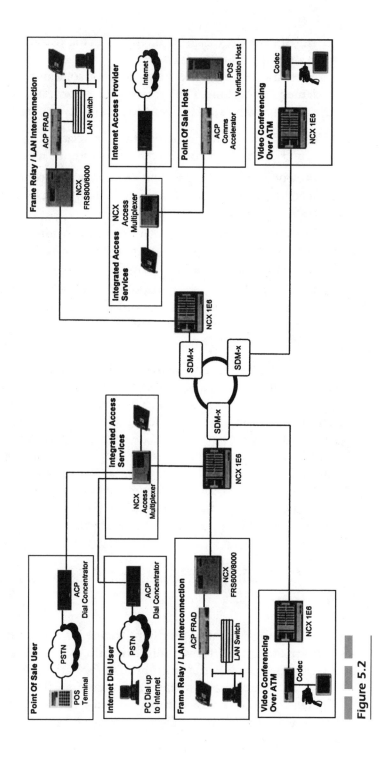

Figure 5.2

125

broadband services, each requiring differing throughput characteristics, over a single network infrastructure.

Telematics enables true ATM network solutions through Subscriber-Oriented Networking. The specific product marketing objectives behind the development of Subscriber-Oriented Networking are:

■ Offer networks that recognize that the relationship between subscriber and network is controlled and managed by SLAs.

■ Provide bandwidth allocation and congestion control mechanisms that are tied to an understanding of each subscriber's SLA and traffic characteristics.

■ Deliver more cost-effective solutions by providing superior SLA/Quality of Service (QoS) protection, while operating networks at the highest possible utilization.

■ Provide multimedia voice, data, and video services that support a broad range of server applications ranging from on-line transaction processing, client/server, and real-time interactive video, to POTS and VPN offerings.

Today's networks deal with bandwidth management and congestion control with standard algorithms that are typically applied to *all* users equally because they happen to be connected to the same link or node. In an ATM environment, where multiple applications with completely different traffic characteristics are being supported, these techniques are ineffective. The network must be capable of providing diverse services, including future multimedia integrated services, as well as supporting the migration of today's services to a common infrastructure.

Subscriber-Oriented Networking solutions allow the networks to guarantee service levels—not just bandwidth.

5.1.7   Making the Most of Expensive Resources

The combination of Subscriber-Oriented Networking and the multi-service statistical networking capability of ATM enables NCX 1E6 networks to deliver the maximum information throughput from the installed bandwidth. In the current competitive environment, where increasing amounts of bandwidth must be purchased at commercial rates, this throughput delivery can make big differences in network costs.

5.1.8 Implementing Subscriber-Oriented Networking

Subscriber-Oriented Networking is based on the principle that each network user will have subscribed to a given level of service from the network, either by explicit contract or by implicit agreement through its access mechanism, location, or identity. It is the function of the network, through its management systems, to meet this agreed level of service at minimum cost to the network operator.

Within the NCX 1E6, a variety of subsystems provide this capability, including metering, congestion management, and resource commitment. Together, they support the concept of Subscriber-Oriented Networking.

At the network's access point, all subscribers are controlled by their agreed traffic profiles through hardware and software systems inherent to NCX 1E6 I/O modules. Standard traffic shaping parameters, such as peak information rate (PIR), sustained information rate (SIR), and cell delay variation (CDV), are used by the ATM services to define and implement this metering activity.

The NCX 1E6 defines a comprehensive set of subscriber/service-related capabilities designed to support the operation of multiple virtual private or public networks. These capabilities include:

■ Masking the physical network addressing schemes from the individual user and implementing addressing schemes that are private to the virtual network

■ Implementing and managing user-specific QoS parameters

■ Providing access and call admission strategies based on each subscriber's SLA

■ Call routing and resource allocation based on the subscriber's and the NSP's requirements

Moreover, the network can determine if it can meet the user's service requirements at connect time and negotiate a lower QoS, if required. This is important from the perspective of the new user and also for the other network users, whose QoS should be protected from overcommitment of resources.

Usage Parameters The use of the network is controlled by a set of parameters that fall into two broad categories: QoS and Grade of Service (GoS). These parameters specify the behavior of the system's user plane (U-plane) and control plane (C-plane), respectively (see Fig. 2.1).

The NCX 1E6 network permits administrators to assign usage parameters in the form of predetermined *profiles* that are engineered by the NSP on a networkwide basis. Profiles can differ in cell transmission priority, node occupancy, cell loss priority, peak information rate, sustained information rate, and maximum cell burst.

5.1.9   Virtual Networks

Virtual networking is the capability of delivering multiple independently managed—or viewed—networks over the same network infrastructure. For the NCX 1E6, this is an inherent element of Subscriber-Oriented Networking, as it allows the network operator to view the service exactly the way the customer perceives it. The capabilities that enable the NCX 1E6 to support virtual networking include

- Management service views
- User private network addressing
- SLAs by user/virtual network
- Service-oriented traffic management capability
- User management views

5.2   NCX 1E6 Architectural Design

5.2.1   Introduction

The NCX 1E6 architecture consists of a central switching fabric connected over internal buses to 16 I/O modules, capable of supporting link speeds from fractional DS0 up to OC-3/STM-1.

The system operates separate adaptation processes interconnected by a common ATM switching fabric. Each I/O module provides a complete switching and adaptation environment, including traffic management and control plane processes, as shown in Fig. 5.3. Traffic between the processes is carried through the common switching fabric.

The NCX 1E6 operating software provides synchronization and management of all system resources.

NCX 1E6 systems use a multiprocessor architecture that supports the SDL design methodology. The close relationship between hardware and

Figure 5.3

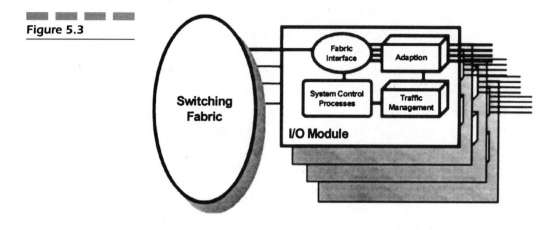

software architectures is key in meeting the very high mean time between failures (MTBF) design goals established for the NCX 1E6.

This tight coupling between architecture and system software is what enables *true* SLA management down to each individual user, for each application supported within the multiservice network.

Figure 5.4 provides an overview of the software components that are described in subsequent sections of this chapter.

U-Plane Implementation   In an ATM switch, the U-plane is responsible for configuring services based on parameters provided by the C-plane, establishing connections based on those parameters, and handling data flows between line interfaces and the switching fabric.

Figure 5.4

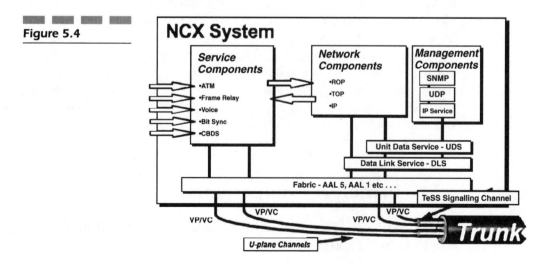

The NCX 1E6 U-plane implementation is highly mechanized and uses custom-developed application-specific integrated circuits (ASICs) to meet product performance and price goals.

The U-plane utilizes three specific ASIC complexes to provide the ATM switching fabric:

■ Transmit (TXGA) and Receive (RXGA) gate arrays responsible for implementing traffic management, such as policing, shaping, and metering

■ Segmentation and reassembly (SAR)

■ Switch-embedded array (SEA)—the switching matrix

C-Plane Implementation The C-plane handles all virtual connection-related functions and also performs the critical functions of addressing and routing. In the NCX 1E6, C-plane implementation benefits greatly from the System Definition Language (SDL) methodologies used in its design. SDL significantly improves the ability to describe the design of new or enhanced functionality to NSPs. SDL also is key to achieving the product's high MTBF targets.

M-Plane Implementation The management plane (M-plane) enables equipment management to be performed through an external workstation running OMS. This workstation can be connected to one or more NCX 1E6 systems through an Ethernet connection.

Buffer Strategy The NCX 1E6 is a nonblocking switch with sufficient backplane bandwidth to ensure that received cells are transferred immediately to the transmit-side IOC. This means that input-side buffering is unnecessary, and cell marking can be efficiently accomplished, resulting in a need to discard only when network trunks are genuinely congested.

Each transmit IOC has buffer space for 8704 cells, allowing short-term trunk congestion to be tolerated. The intelligent queue management ensures that traffic is handled appropriately relative to its delay tolerance and that, should it become necessary to discard cells, noncompliant "low-impact" cells are discarded first.

5.2.2 Principal Components

Within the NCX 1E6 there are three principal functional subsystems: switching fabric, system control processors, and I/O modules.

Switching Fabric   The switching fabric is the heart of the hardware. Designed exclusively to switch ATM cells, it provides the switching capability of the system, taking input cells from the I/O cards and delivering them to their appropriate destination I/O cards.

The switching fabric provides switching performance in excess of 5 million cells per second with a continuous 3.5-Gbps throughput. Its capacity exceeds that of the I/O cards, ensuring a nonblocking environment.

System Control Processor   The system control processor provides an environment for operating nodal and network control plane functions. This processor resides with the switching fabric on the ATM control switch (ACS) cards and utilizes the switching fabric to communicate with I/O modules.

I/O Modules   I/O modules provide physical, electrical, and protocol interfaces. They are physically constructed in two parts to improve reliability and mean time to repair (MTTR)—an active main card and passive transition subcard assemblies.

The I/O modules are responsible for the segmentation and reassembly of data to and from the internal ATM cell format utilized within the system, as well as traffic policing, metering, and shaping. Each I/O module has an aggregate throughput of 160 Mbps, allowing the direct connection of OC3 and STM-1 links to the NCX 1E6.

Each IOC also contains a control processor that supports link- and call-related software components.

5.3   Hardware Overview

5.3.1   Chassis

The NCX 1E6 single chassis mounts in a 19-in rack and occupies 12 rack units (RUs). Each RU is equivalent to 1.75 in. Slots are available for power supplies, ACS cards, and input/output cards (IOCs).

Common Cards   The chassis contains the following common cards:

- Power Supply 1—active
- Power Supply 2—optional hot standby (not shown)

- ACS 1—active
- ACS 2—optional standby

Line Cards The chassis can accommodate up to 16 line cards:

- IOC (1 through 16)—configured as active and optional standby

Figure 5.5 provides a graphical representation of the front panel and physical card positions.

Midplane All cards are connected to a midplane that traverses the chassis width. The NCX 1E6 uses the midplane, rather than a backplane, to separate the physical line connections from the I/O modules. Physical lines are connected to transition cards on the opposite side of the midplane from each associated line card. Transition cards have few active components, ensuring long MTBF. This design allows I/O modules to be removed without disturbing line-side connections, thus reducing the possibility of operator-induced faults.

Physical Connections The NCX 1E6 rear panel appearance varies depending on model type and configuration.

Figure 5.6 shows one particular NCX 1E6 model configured with twelve T1/E1 ports, two T3/E3 ports, and three SONET/STM-1 ports.

Figure 5.5

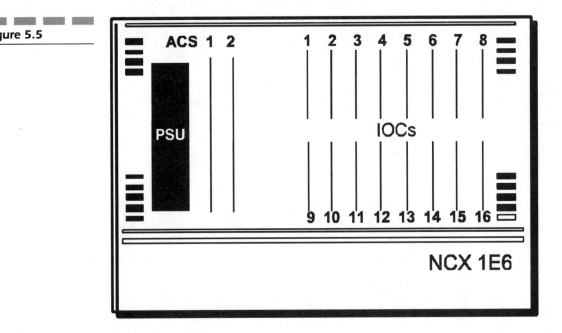

Figure 5.6

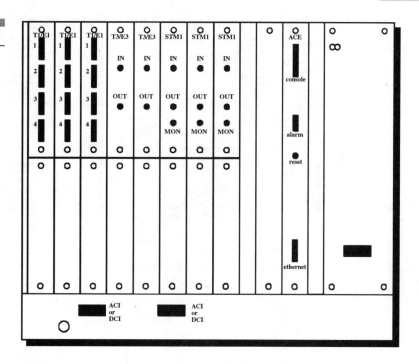

5.3.2 Power Supply

The NCX 1E6 can be operated with AC or DC power, and has autoranging power supplies to meet worldwide power supply standards.

AC power supplies operate from 100 V AC to 250 V AC. DC power supplies operate from −48 V DC to −75 V DC.

Additionally, the NCX 1E6 can be equipped with a second power supply to provide nonstop operation in the event of primary power supply failure.

Supplied from separate main power sources, both power supplies use diode-buffered load sharing to ensure a seamless cutover in the event of failure, as well as substantially increasing the life of the power supplies by reducing stress levels under normal operating conditions.

5.3.3 ATM Control Switch

The NCX 1E6 uses a proprietary switch architecture to meet cost/performance goals. It provides switching and adaptation functionality in the T1/E1 to OC-3/STM-1 range.

Figure 5.7

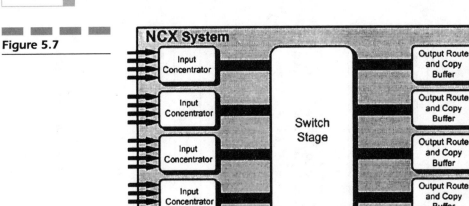

An overview of the ATM control switch architecture is shown in Fig. 5.7. Line cards interface directly to this module through Transmit (TX) and Receive (RX) buses on the system midplane.

The ACS contains three discrete stages:

- Input stage
- Switching stage
- Output stage

Input Stage The input side of the ACS and the receive side of each IOC are interconnected by a point-to-point bus to transmit information from the IOC to the ACS.

In the NCX 1E6, each IOC is connected to the ACS through a 4-bit-wide, 44-MHz RX bus, enabling a data transfer rate of 176 Mbps. Information is carried on each RX bus in a proprietary cell format. Each cell, known as a backplane cell, contains 5 bytes of ATM header, a 48-byte payload, and an additional 3 bytes of NCX 1E6 proprietary data.

Switching Stage The switching stage takes backplane cells from the input RX buses and places them on a time-shared bus. This is done in a round-robin, nonstatistical manner. The bandwidth of this stage is greater than the sum of the input space to ensure that the switch is non-blocking. This "time" stage has an 80-bit-wide bus operating at 44 MHz, with some guard time between consecutive cells on the time bus, enabling a cell switching rate of 3.52 Gbps.

Output Stage   The output side of the ACS and the transmit side of each IOC also are interconnected by discrete buses. The NCX 1E6 has 16 IOC slots; hence there are 16 IOC TX or transmit buses. Each TX bus is 16 bits wide and operates at 44 MHz.

The information received on these buses is still in the NCX 1E6 backplane cell format. The placement of cells on these TX buses is slaved to the ACS, which drives a central cell timing signal to all modules. While a cell is present on the time bus, all output space stages perform an address match on the NCX 1E6 cell header.

If the bit corresponding to a given output space is set, that output bus copies the cell from the time bus into a local first-in first-out (FIFO) buffer associated with that TX bus. Multiple output spaces can receive a copy of the cell from the time bus, providing the foundation for accomplishing a first level of multicasting.

Since the bandwidth of the time bus is greater than the bandwidth of the TX bus, a FIFO buffer is needed to accommodate instantaneous burst conditions. This FIFO buffer is 64 cells deep, and introduces a worst-case delay of approximately 60 µs. The output stage can transfer data at the rate of 704 Mbps.

5.3.4   I/O Cards

The NCX 1E6 can support up to 16 IOCs that communicate with each other through the ACS and its buses.

Physical Interfaces   The NCX 1E6 has a modular design that allows many different IOCs to be developed, providing support for all types of data, including video, voice, and LAN traffic. There are a number of IOC variants providing support for the following physical interfaces:

- STM-1 (G.703 75-Ω coax)
- STM-1 (G.957 optical)
- SONET OC-3c (G.957 optical)
- E3 (G.703 75-Ω coax)
- T3 (G.703 75-Ω coax)
- E1 (G.703/G.704—120-Ω twisted pair or 75-Ω coax)
- T1 (100-Ω twisted pair)

IOC Configurations   IOC configurations are governed by the number of physical slots available and the constraints of product characterization. The NCX 1E6 can support a combination of up to 16 STM-1/OC-3 or T3/E3 access or trunk IOCs. A maximum number of 64 T1/E1 ports is supported.

Figure 5.8 shows a general IOC model. Note that some IOCs will have only a single port—a single *physical* subscriber interface—not four as shown in this particular model.

Two general types of IOC are specified for the NCX 1E6—cell-based modules and non-cell-based modules. This section provides a brief description of their capabilities and implementation.

Cell-based Module   Cell-based modules receive 53-byte cell traffic on an input port and, after validating the input cell header, pass it directly to the RXGA for processing by the ACS.

Non-Cell-based Module   Non-cell-based modules receive frame-based or circuit emulation traffic at an input port. The initial function performed on this traffic is the conversion from its native protocol to ATM cells through the adaptation process.

Once in cell format, the cell is presented to the RXGA, which performs policing, routing, and header manipulation to convert it into an

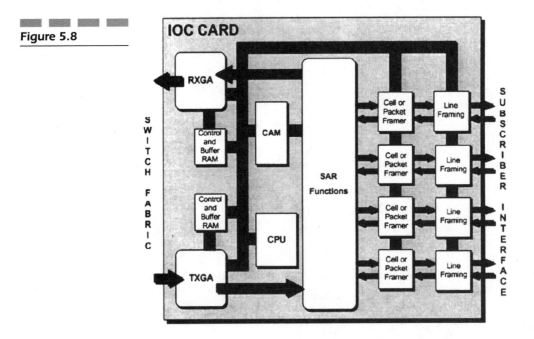

Figure 5.8

ATM cell. Once converted, it is transmitted on the RX bus to the switch.

Non-cell-based modules also receive ATM cells on their TX bus and, after validating the input cell header in the TXGA, perform class servicing, ATM header removal, and multicasting if necessary. They then pass the cell to the conversion complex, which reassembles the original data in their native format.

Adaptation Support   IOCs and ACSs each contain a reduced instruction set computing (RISC) processor that provides a common processing environment—the Communications Executive (CEE). All software processes within the NCX 1E6 operate within this CEE environment. Each specific protocol adaptation is implemented using the appropriate software modules running on the IOC. For instance, the T1/E1 rate modules can accommodate many services, such as frame relay, bitsync transport, and ATM.

These subscriber interfaces are described in Sec. 5.3.5.

System Resilience   The NCX 1E6 is designed to be configured as a fully redundant switch.

The architecture allows I/O, power and cooling systems, and standby switching modules to be replaced in live systems.

The following modules can be provided for redundancy:

- Power supply
- ACS
- IOC

The 16 I/O slots are for any mix of wideband four-port cards or broadband single-port cards. Redundancy for broadband cards is 1:1, with the redundant card immediately to the right of the main card. Changeover to a standby module is fully automatic on detection of failure.

The IOCs can be configured as nonredundant or as 1:1 or 1:*N* redundant, depending on card type. NCX 1E6s normally are delivered with dual switch fabric subsystems providing 1:1 redundancy, although this is optional. The power module also can be fully duplicated.

The NCX 1E6 is designed to run continuously in the event of a fan failure. Fan assemblies can be hot swapped without affecting subscriber services.

Diagnostic Features   Alarms, or events, are generated by the NCX 1E6 when a fault or failure of any module, cross-connect, or transmission medium is detected. The nodes automatically update the network topology to ensure continued service to subscribers. Alarms are transmitted to the Network Management Center, indicating the type of failure, possible causes, and corrective action.

A range of loopback, test, and diagnostic features are provided depending on the interface and service used.

Recovery from Failures The system architecture ensures that a power supply, switching fabric, or individual IOC failure will have no impact on system performance, provided that the system is equipped with appropriate redundancy options.

The NCX 1E6's very high level of reliability and resilience ensures that an individual node will maintain QoS in the event of system component, power or cooling system, or IOC failure. Each system automatically reallocates services to redundant components without operator or management intervention.

Subscriber interfaces provide physical connection, electrical connection, and mapping of communications protocols—native ATM and non-ATM—into the NCX 1E6, enabling voice, data, and video support over an ATM infrastructure.

5.3.5 Subscriber Interfaces

The NCX 1E6 supports the following subscriber interfaces:

- ATM
- Circuit emulation
- Bitsync
- Frame relay
- CBDS/SMDS
- Digital voice/PABX

ATM The ATM UNI access function performs the following functions related to the Physical Layer:

- ATM cell handling—delineation, scrambling, and header error control (HEC) processing (error detection and correction)
- Mapping ATM cells into appropriate payloads
- Operations, administration, and maintenance (OAM) flows (optional)

The access function practically implements all B-ISDN functionality needed at this reference point and the mapping of ATM cells into synchronous frames, i.e., G.709 or G.804. ATM cell mapping is compulsory, since the switching occurs at the cell level. Cells must be extracted from the payload before switching and mapped back into the payload after switching.

OC-3/STM-1 ATM Interface Card   The OC-3/STM-1 ATM IOC provides effective full throughput of ATM user cells at rates up to 155 Mbps. Both electrical (G.703/12) and optical (G.957) interfaced versions of the STM-1 card are available.

F1 to F5 OAM flows are handled as defined in I.610 for SDH-based signals, European Technical Standard (ETS) 300-300.

T3/E3 ATM Interface Card   The ATM T3/E3 IOC transmits and receives cells at 45 or 34 Mbps, respectively. The IOC complies with the electrical interface specifications according to ITU-T Recommendation G.703, as well as the standard ATM UNI specification. Standard ATM cells are mapped into a G.804/G.832 frame. Once synchronization has been achieved with respect to ITU-T Recommendation G.804, cell delineation is performed.

F1 to F5 OAM flows are handled as defined in I.610.

T1/E1 ATM Interface Card   The T1/E1 ATM IOC provides standard ATM UNI cell transmission/receipt at 1.544 or 2.048 Mbps, respectively. The card complies with the electrical interface specifications according to ITU-T Recommendation G.703, as well as the standard ATM UNI specification. Standard ATM cells are mapped into a G.804 frame. Once synchronization has been achieved with respect to ITU-T Recommendation G.804, cell delineation is performed.

F1 to F5 OAM flows are handled as defined in I.610.

Circuit Emulation   The NCX 1E6's circuit emulation IOCs adapt unstructured and structured framed signals into ATM cells for switching across an ATM infrastructure. This capability allows the NCX 1E6 to switch information streams that are sensitive to clock synchronization, such as voice or video, as native terminations on the NSPs edge-switching fabric.

Unstructured Circuit Emulation   In the G.703 synchronous mode, the signal will be locked to the synchronization clock. AAL1 will operate in synchronous mode without invoking any rate adaptation mechanism.

If the network is not using a single-source synchronization clock—plesiochronous—a controlled slip rate will occur as recommended in ITU-T Recommendation G.822.

The user information bandwidth in a virtual channel connection (VCC) will be constant 1.544 Mbps or 2.048 Mbps, as appropriate. The delay variation buffer length is a function of the maximum cell delay variation (CDV) and a tradeoff between the total transfer delay and the slip rate. For voice services, where the 8-kHz sampling rate has to be retained, it is recommended that 125-µs controlled slips be provided, i.e., 256 bits. To lower the delay for other services, 1-bit slips can be used. In such cases the slip rate will be increased.

In the G.703 asynchronous mode, the signal rate at the 1.544/2.048-Mbps output is recoverable from the cell stream by the synchronous residual time stamp (SRTS) or adaptive mechanism.

Adaptive clocking reproduces the source clock independently of any network synchronization signal. This is accomplished by monitoring the variation in receive cell buffer depth and adjusting the clock to keep a constant buffer depth. Therefore, it is "synchronized" with the transmitter clock. There is a compromise between stability and delay. As delay increases, receiver phase lock loop integration time takes longer, resulting in a more stable clock.

SRTS timing relies on the presence of a synchronized network timing signal, such as that required for synchronous mode. The difference between the phase of that clock and the phase of the transmitter clock is measured on entry to the network. A measure of the difference is passed across the network as part of the AAL1 sequence byte, and is used to regenerate the clock at the receive side.

For services where the jitter/wander specification allowed is less stringent, the adaptive method is recommended. In such a case, there is no need to have the same clock frequency at the send and receive ends.

Structured Circuit Emulation   Structured circuit emulation will be provided in synchronous mode using AAL1 and the SDT 8-kHz structure. If the network is not using a single-source synchronization clock (plesiochronous clock), a controlled slip rate will occur as recommended in ITU-T Recommendation G.822.

Only the active bytes will be transported over the VCC. Therefore, the required user-information bandwidth in a VCC depends on the number of active time slots in the G.704 frame ($n \times 64$). It should be noted that for a low value of n, when carrying voice traffic, echo cancellation may be required. Depending on the QoS objectives of the G.704 services carrying voice, the NSP may specify a minimum value of n. The G.704 service provided supports this capability. However, it has a range of n from 1 to 31 in a general case.

The transport of Time Slot Zero (TS0) in the VCC is subject to the required applications needed by the NSP. If only the remote alarm indication (RAI) needs to be supported, there is no need to occupy an entire time slot in the VCC. If other applications (e.g., 4 kbps data link) have to be supported, the TS0 information can be transported transparently over the VCC together with the other $n \times 64$ kbps time slot information. Alternatively, a separate signaling cell stream can be used, where appropriate information from TS0 is extracted at the ingress point and delivered to the egress point. The latter method allows the NSP a clear evolution path to multipoint G.704 services.

The general specifications will comply with the following recommendations. See the system specifications section for other applicable standards.

- Jitter/wander: ITU-T Recommendation G.823
- Slip rate: ITU-T Recommendation G.822
- Error performance: ITU-T Recommendation G.826
- AIS and LOS criteria: ITU-T Recommendation G.775

Clock Distribution   For an effective high-quality circuit emulation service, great care is needed in clock distribution planning. The NCX 1E6 has a clock management capability that enables clock signals to be distributed through the network. Increasingly distributed, highly accurate clock sources are being used. These sources, such as GPS, also can be used by the NCX 1E6 for synchronization. ANSI T1.101 and ETSI 300 462 describe the synchronization requirements for synchronous transmission networks.

T1/E1 Circuit Emulation Interface Cards   The T1/E1 circuit emulation IOCs support G.703 unstructured and G.704 framed signals for 1.544 and 2.048 Mbps, respectively.

Bitsync   The NCX 1E6 supports a bitsync transport mechanism that provides a means of supporting HDLC traffic such as X.25 and SDLC/SNA data across an NCX 1E6 ATM network.

X.25 logical channels and SNA LU-LU sessions are mapped 1:1 to an ATM VP/VC (see Fig. 5.9).

Figure 5.9

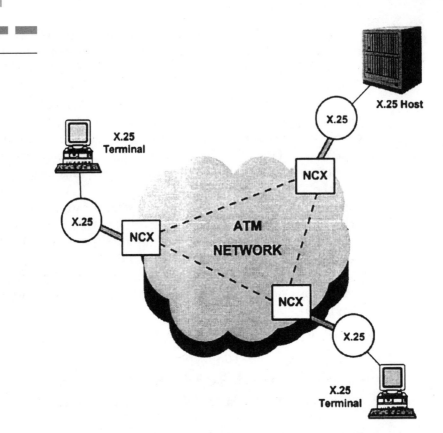

Bitsync allows the transmission of this type of data more efficiently than can be achieved using circuit emulation. All idle flags are removed from the data prior to transport, thus allowing the statistical gain of a variable bit rate (VBR) service to be provided for any bitsync protocol. This capability to support legacy protocols within an ATM multimedia environment provides flexibility in providing outsourced network solutions.

T1/E1 Bitsync Interface Cards   The T1/E1 bitsync/frame relay IOCs utilize AAL5 to map information streams to ATM cells through $N \times M \times 64$-kpbs full channelized interfaces.

Frame Relay   The frame relay (UNI) over ATM implementation on the NCX 1E6 supports permanent virtual circuit (PVC) services using AAL5. The NCX 1E6 architecture is also designed to support frame relay NNI and switched virtual circuit (SVC) services. At the physical level, the bit

Figure 5.10

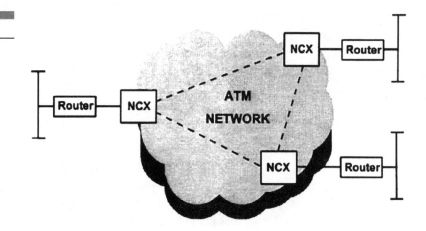

rate is from 64 kbps to 1.544 Mbps or 2.048 Mbps. Effective throughput of ATM user cells is up to 2.048 Mbps. Frame sizes of up to 4096 bytes are supported.

All intelligent protocols such as frame relay and ATM are terminated on the IOC (Fig. 5.10). Only relevant data are transferred across the network. LMI and call processing are handled at the network point of entry. The out-of-band intranetwork call processing capabilities of the NCX 1E6 are utilized to establish the circuit across the network.

Frame Relay Congestion—Avoidance and Recovery   Within any frame relay network, a critical function of the network is to first avoid but, if required, manage network congestion. Congestion control is vital in preventing the network from suffering failures through oversubscribing resources.

The NCX 1E6 employs several sophisticated congestion control mechanisms for avoidance and recovery. This provides high network resiliency and allows an NSP competitive flexibility in providing differing levels of QoS.

Congestion is indicated to the access community by setting the forward and backward explicit congestion notification (FECN and BECN) bits within the frame relay header. If a circuit experiences network congestion, the FECN bit is set within the outgoing frame to indicate potential network congestion. The originating and terminating nodes maintain a BECN bit that is used to reflect the incoming FECN bit. If a frame arrives for transmission to an access device with the FECN bit set, the BECN bit will be set in the next outgoing frame from the access device to the network.

Once frame relay frames have been passed to the NCX 1E6 for processing, internal congestion mechanisms are used.

The following frame relay parameters are mapped directly to ATM:

- FECN to EFCI
- D/E to CLP
- C/R to CPCS-UU
- DLCI to VP/VC

The NCX 1E6 implements ANSI T1.617 with respect to frame input within a frame relay node to provide and maintain desired QoS levels. There is an available buffer pool for a circuit that corresponds to the committed and excess data rates for the circuit. Frame relay buffering is in addition to the standard ATM buffering available in the system. The NCX 1E6 is equipped with the same traffic and congestion control management algorithms inherent to NCX FRS switches, enabling consistent, powerful frame traffic management across NCX cell and frame switches. As a frame arrives for a circuit, the available buffer space is incremented, up to the maximum, in proportion to the time since the previous frame's arrival. The frame size then is compared to the available buffer space to determine if the frame should be forwarded, marked DE and forwarded, or discarded. The strategy for forwarding, marking DE, and discarding is based on the following parameters in accordance with standards recommendations:

- Circuit settings of the committed information rate (CIR)
- Excess information rate (EIR)—the maximum rate at which the network agrees to transfer data under any conditions
- Bc—committed burst size
- Be—excess burst size

The effectiveness of this congestion avoidance depends on the connected devices acknowledging the FECN and BECN notifications and reducing their traffic.

T1/E1 Frame Relay Interface Cards The T1/E1 bitsync/frame relay IOCs utilize AAL5 to map information streams to ATM cells through $N \times M \times 64$-kbps full channelized interfaces.

CBDS/SMDS The Connectionless Broadband Data Service/Switched Multimegabit Data Service (CBDS/SMDS) interface is intended to provide a connectionless data service over an ATM network. CBDS is intended to operate at rates up to 45/34 Mbps, but many applications may not require the full T3/E3 rate.

The CBDS/SMDS interface enables the interconnection of SMDS MANs over a B-ISDN ATM network. This eliminates the need for a specific dedicated network for interconnection of MANs, such as a network of MAN switching systems (MSS).

The following information describes the initial SMDS connectivity capability of the NCX 1E6. Native SMDS terminations are under consideration for future NCX 1E6 releases.

Initially, the SMDS DXI can be implemented with an external DSU that will be connected to an ATM UNI at 45/34 Mbps. This configuration will supply all the functionality required for CBDS, with the exception of the CLNAP layer. If the external DSU has an ATM DXI, then this DXI can be used as well (see Fig. 5.11).

Digital Voice/PABX   Digital voice connectivity on the NCX 1E6 is supported, ranging from a small number of analog voice ports at a branch office to hundreds of voice circuits at a central office. The initial G.703 circuit emulation capability is available as part of the basic feature set of the NCX 1E6.

Digital Voice at T1/E1 Speeds   Digital voice interfaces will be supported by the T1/E1 circuit emulation IOCs. Once a port is configured to offer a PBX interconnection service, the following options are available:

- Transparent 1.544 or 2 Mbps pipe for PABXs conforming to ANSI T1 or G.703 specifications
- Transparent passage of all framing and signaling time slots
- Multipoint services through circuit emulation–structured data transfer (CE-SDT)

Figure 5.11

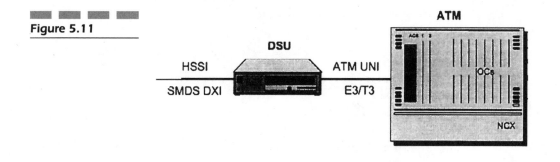

- Independent routing of each time slot—or group of time slots—to a separate destination through translation of time slot 16 information (through CE-SDT)

- E1 connections conforming to G.703, G.704 (through CE-SDT), and G.732

- Analog-to-digital voice applications using the NCX access multiplexer

The digital interface provides 24 time slots for T1/D4 framing or 32 time slots for E1/G.704/G.732 framing. Each circuit emulation IOC port can carry voice or data time slots independently. For E1 rates, time slot 16 can be used for signaling or as a standard voice/data channel.

Compression The NCX 1E6 supports voice compression through the QC300 dedicated compression module attached to any G.703/G.704 port. Compression is based on the 16-kbps low-delay, code-excited linear prediction (LD-CELP) standards (G.728).

Voice channels from the PBX interface are routed through the compression equipment, which is configurable on a per-channel—time slot—basis.

DSP implementations of data/speech discrimination algorithms and fax demodulation and modulation also have been developed. Fax demodulation allows 4:1 compression of fax signals. Using the above technologies, Telematics' systems are able to treat voice and voice-band data (VBD) signals independently, achieving compression on both.

The QC300 enables this solution to be used with the 16-kbps LD-CELP compressor. The voice compression and fax demodulation DSP solution is integrated in a compression module that is able to receive individual streams and compress voice and fax signals. A QC300 module also is used at the far end for decompression. Non-fax VBD signals are sent transparently without compression. Telematics also provides voice detection capabilities through the DCME product line, enabling digital speech interpolation (DSI) technology to further compress the voice signals using VBR voice transmission.

5.4 Telematics Network Architecture

The Telematics Network Architecture (TNA) is a key part of the Subscriber-Oriented Networking philosophy.

TNA contains eight architectural submodules. Each submodule describes the responsibility and behavior of a functional area of an NCX 1E6 network. As such, users can understand and specify requirements that will be consistently interpreted across NCX 1E6 ATM switching platforms. It is this architectural view and functional breakdown that leads to the characterization of the NCX 1E6 as a Subscriber-Oriented Network solution. The architectural submodules are as follows:

- Connection
- Call
- Subscriber
- Network
- Node
- Topology
- Route
- Media

5.4.1   Connection

The connection submodule establishes connections between any two points at the network boundary in support of a call between two subscribers. Connections can be customized to provide the method of data transfer most suitable for the subscriber's applications, at a cost established by the network administrator.

All connections in the network transfer data in accordance with a QoS profile derived from a record of the subscriber's relationship with the NSP. This profile is the subscriber's SLA.

5.4.2   Call

A call is the basic unit of service that the network offers its subscribers. A call adds value to the transfer capability provided by a connection by coordinating the use of two or more connections over a period of time to provide a duplex or multicast communication path between two or more subscribers.

In the case of a duplex or multicast call, the NCX 1E6 routes each of the connections individually so that the best route is obtained for traffic in each direction. Each connection has its own QoS parameters, so asymmetric traffic flows can be accommodated.

Calls have QoS parameters that are derived from the subscriber's SLA. These parameters specify the extent to which the call will recover from any network problems within the network. For example, the NCX 1E6 can reestablish connections in the case of a network failure.

5.4.3   Subscriber

Subscribers are the network's customers. Any user of network resources is, by definition, a subscriber. This concept extends to management and signaling traffic as well. Subscribers negotiate service terms with the NSP. The result of this negotiation is defined in an SLA, which records not only the extent to which a subscriber is permitted to consume network resources, but also the obligations that the network has to that subscriber.

When a subscriber first connects to the network, the node where the connection occurs refers to the subscriber's SLA to characterize hardware and software in line with the subscriber's requirements. This is key to providing virtual networks and nomadic subscribers. The process differs from traditional packet-switching equipment, where the configuration of a node is driven by the appearance of a particular subscriber, and can be thought of as "configuration on demand."

Because an NSP needs to refer only to the central "master" copy of an SLA, and because the distribution of subscriber information is managed automatically by the network, the cost to establish and manage subscriber information is significantly reduced.

The SLA contains not only the QoS parameters required by call and connection, but also subscriber-related GoS parameters, such as the time of day a subscriber is permitted to use the network and which subscribers are permitted to place calls.

Also, the SLA specifies how subscribers' calls are established within the network. In addition to specifying the terms under which a subscriber can establish temporary calls, the SLA also can specify that a "permanent" group of calls be established by the network on behalf of the subscriber whenever it is connected to the network. This is the means by which PVCs are implemented and ensures that when a subscriber is not present on the network, all resources are made available for use by other subscribers.

Subscribers' SLAs are maintained at the management center and at the node where the subscriber currently resides.

5.4.4 Network

The TNA network submodule defines how NCX 1E6 software services operate as a distributed application, enabling larger, overall network utilization objectives to be achieved by the NSP.

The NCX 1E6 uses standard software interfaces into UNIX computing environments so that conventional data-processing techniques can be used to modify and characterize the NCX 1E6's behavior. The management of the NCX 1E6 also falls into this category. SNMP object definitions and APIs provide compatibility with standard third-party applications and management systems.

5.4.5 Node

The TNA node submodule describes the basis of how individual devices participate in the network, and how resources provided by different devices are coordinated and managed. Each time a connection is established, the resources used for that connection are calculated, allowing an overall costing strategy to be executed by the NSP.

The ability to support nomadic subscribers and to allow devices to be configured "on demand" is addressed by the node submodule. The partitioning of node resources into "virtual nodes" allows support of a virtual network concept.

By building virtual networks from virtual nodes, rather than just as a management concept built on physical devices, a greater degree of freedom is offered to the NSP in terms of sharing resources while protecting every SLA.

5.4.6 Topology

The topology submodule defines how the NCX 1E6 can cooperate in distributing and maintaining information relative to network topology. The topology submodule is key to reducing the cost of owning the network, and provides self-learning phases of establishing basic connectivity.

5.4.7 Route

The route submodule describes how resource sets are selected to participate in a given call. The NCX 1E6 routing strategy is discussed in Sec. 5.6.

5.4.8 Media

The media submodule establishes how different transmission media can be used in an NCX 1E6 network.

High-level call signaling procedures can be used to establish a connection across subnetworks. That connection then can be used as a single medium within the overall NCX 1E6 network. This provides the ability for the NCX 1E6 to use a range of bearer subnetworks as "virtual trunks"—essential in supporting bandwidth on demand services across bearer subnetworks.

5.5 Telematics Extended Signaling System

Telematics Extended Signaling System (TeSS) is the NCX 1E6's internode signaling system. TeSS is a proprietary signaling protocol based on Q.2931 signaling standards. Each element of TeSS performs a specific function, communicating on a peer-to-peer basis with each logically adjacent element.

The major elements of TeSS are:

- Internet Protocol Service
- Topology Provider
- Route Provider
- Central Call Service

These elements are responsible for route selection, call setup and reconnection, traffic rerouting during node or network facility failures, and circulating network management information between nodes.

The relationship among the various TeSS elements and other software components is shown in Fig. 5.12.

TeSS uses the ATM Layer for transmission and a transparent data link service at layer 2 to provide topology, routing, call services, and UDP/IP network management transport mechanisms.

TeSS supports the following functions: semipermanent VCs, switched VCs, peak cell rate allocation, and statistical allocation. Each VC is assigned to an SLA definition that determines the traffic management, shaping, and QoS parameters to be applied to that individual VC.

Figure 5.12

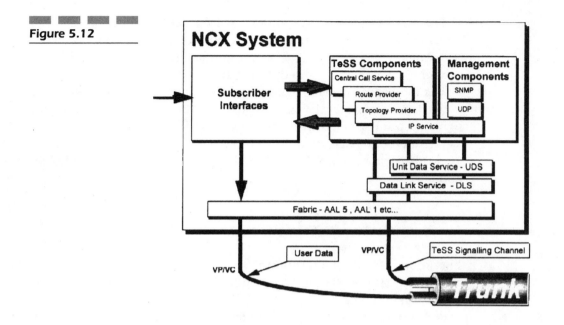

5.5.1 Internet Protocol Service

The Internet Protocol (IP) Service is designed to

■ Enable NCX 1E6 nodes to be part of an IP management network.

■ Provide low-level protocols to support node startup and bootstrap.

■ Support unit data service (UDS) and data link service (DLS) for conveying datagrams across NCX 1E6 networks.

■ Provide a UDP user-data size of up to 4000 octets.

■ Support Routing Information Protocol (RIP) for automatic route table update when the IP topology changes, reducing the amount of explicit configuration required.

The IP Service is not a subscriber service, but exists in support of signaling and management traffic.

5.5.2 Topology Provider

Topology Provider (TOP) enables NCX 1E6 nodes to *self-learn* the network topology in which they operate.

The role of the topology process is to generate a map of all other nodes by establishing the node address (NID) of adjacent nodes, then distributing this information to adjacent nodes. These nodes then will relay this topology information to their neighbors.

The topology process is a fundamental requirement of any NCX 1E6 network, as it provides the essential information required for routing of signaling and U-plane connectivity networkwide.

The topology database and route mechanisms also provide support for load sharing on a VC basis across multiple links between adjacent nodes.

5.5.3   Route Provider

The Route Provider (ROP) routes connections between *subscribers* in a manner that supports their QoS requirements, while optimizing the use of network resources.

ROP utilizes a two-phase route determination process that provides effective route selection, while minimizing network overhead using the following criteria:

Which resources have the *potential* to support the requested connection

Which specific resources *should be used* to support the connection requested

When making a call, ROP determines an initial source node route, based on the latest network utilization information. The Central Call Service (CCS) then will attempt a connection. However, should CCS be unable to allocate the necessary networkwide resources with this initial selection, it will ask TOP for an alternative route that will avoid congestion.

5.5.4   Central Call Service

The CCS is responsible for binding together all the network resources required to support an individual call.

The NCX 1E6 implements calls through permanent, semipermanent, and switched virtual connections for both VPs and VCs. VP and VC connections are used to establish circuit emulation and data services.

In addition, CCS is designed to

- Provide a structured architecture that allows extended services to be added.
- Provide services that are structurally compatible with the requirements of broadband networks to establish U-plane connections capable of supporting VBR and CBR transmission.
- Support out-of-band call signaling procedures.
- Support a call QoS that reflects the requirements of subscribers' facilities, route selection, topology capability, and media capacity.
- Support independent routing of the forward and backward connections across the network when configured for simplex operation.
- Support appropriate error mechanisms to allow the network to recover from failure conditions.
- Support the ability to reconnect intranetwork call connections after line failure in connection establishment and data transmission states. A reconnected call connection must satisfy or exceed the minimum QoS requirements of the original connection. Procedures also ensure that usage-related information is carried over from the original to the new call connection in the event of a reconnect.
- Allow establishment and management of "virtual trunks" between any two network nodes so that the trunk can be used as network media, logically connecting two points.
- Allow establishment of call connections and virtual trunks between nodes in different network domains, and support the reconnection of cross-domain call connections at the inter- and intradomain levels.
- Support call usage records compatible with TNA billing requirements.

5.6 Routing Mechanisms

The NCX 1E6 employs source node routing. When a connection is established, the node that originates the request operates as the "master" and manages the entire connection establishment process, treating intermediate nodes in the route as "slaves."

The master node then determines what other nodes in the network are able to support the connection by examining a *topology* database. This topology database resides in every network node and is constructed and maintained automatically from information broadcast from other nodes as their operational state changes.

Based on topology database information and additional information supplied by OMS when the node is deployed, the master node selects one of a set of "potential routes." Here each route is a collection of intermediate nodes and media that has the capability of providing required bearer service attributes.

Having selected a route, the master node then sends a connection request message to each node along the chosen route over the signaling network, requesting it to establish a connection based on the route and bearer service attributes.

Should any node in the selected route be unable to support the connection, the initial route is considered to be invalid and the master node will select an alternative route.

5.6.1   Routing Detail

The NCX 1E6 routes connections between subscribers in a manner supportive of the subscribers' QoS requirements, while at the same time optimizing the use of the network equipment. This is in keeping with a key NCX 1E6 edge service switch objective. The objective is to provide quantifiable service levels to each of its subscribers, but at a controllable cost to the NSP.

In order to perform this effectively, information about *all* network resources needs to be considered when establishing a connection on behalf of a subscriber. Traditionally such centralized approaches to resource management have proved suboptimal because of the data volume that needs to be centralized and the delay incurred by all nodes interacting with the central resource manager.

The NCX 1E6's routing philosophy is based on the need for two types of routing information:

■ Capability

■ Capacity

Capability information is qualitative in nature and is used to determine those resources that have the *attributes* to support a particular type of connection. Examples of capability information are a resource's ability to support isochronous traffic or its ability to dynamically alter its media bandwidth, as with a link or a node. Capability information is relatively static.

Capacity information is quantitative in nature and is used to determine which specific resources are *available* for a particular connection. Examples of capacity information are the utilization of a resource, such as sustainable cell rate on a link, or the number of links to a given adjacent node. Capacity information is relatively dynamic, as it is a function of network traffic load at a given point in time.

Both types of information need to be analyzed in determining the optimum route for a call with a particular QoS. Other than in very small networks, the distribution of such large amounts of data can become a complex and inefficient operation.

The NCX 1E6, however, can synchronize this valuable operating information across all network nodes in near real time. It accomplishes this by maintaining only capability information in the topology database. This allows each node to have an up-to-date view of the network's overall capability and permits the distribution of a "central" route determination strategy. To complement this, each NCX 1E6 maintains capacity information about the resources it directly manages. This information is accessed only when that particular node is requested to commit resources to a connection.

To support the distributed processes necessary to implement the NCX 1E6's QoS and network utilization management capabilities, a separate signaling network is provided. This signaling network uses U-plane connections on each intranetwork link to provide higher-level services between processes in each NCX 1E6.

The routing process remains static as long as there are no changes to the network topology. If the network topology changes, then the routing tables are updated to reflect the network resources available.

5.6.2   Network Failure

In the event of a network failure, the various components of an NCX 1E6 network will generate events depending on the type of failure. There are many types of failures, including failure of the transmission resources, loss of synchronization, QoS parameters not being met, or the inability of bearer services to deliver the level of service required.

Once an event is generated, notification of that event is propagated to the call management function processes for the affected paths and virtual circuits. The call management functions within each ATM system will automatically reestablish the failed paths and virtual circuits, using the updated network topology to reroute the data, ensuring that the correct QoS is provided for each affected subscriber at a minimum resource cost to the network.

As network resources become available again, the automated topology learning process enables paths and virtual circuits to be reestablished through the original transmission path.

Self-learning network topology algorithms are used to ensure that, in the event of node failure resulting from loss of main power, fire, or flood, services can be quickly restored. The network automatically updates its topology, rerouting VPs and VCs around the failure. Following network node recovery or the addition of new nodes or links, network topology is updated to reflect restored or newly available capabilities and capacity.

5.7   Traffic Management

In order to have a system capable of performing at desired levels, certain fixed functionality has been designed into silicon components within the NCX 1E6. This fixed functionality supports the range of input traffic profiles and the QoS required for these profiles. This section provides an overview of the U-plane algorithms and how they support the QoS parameters.

All traffic flowing to a Receive bus can be subject to policing. This policing process is performed following segmentation, and may not be required for all sources, such as a CBR cell flow.

5.7.1   Policing

When a channel is found to be nonconforming, traffic can be discarded or marked. Otherwise, it is sent "as is."

This output marking information is conveyed to other system components to allow discrimination against nonconforming traffic flows, which may include traffic that is not subjected to any traffic shaping in the presence of significant contention for system resources.

The RXGA is the component on an IOC card that normalizes all traffic arriving on a module into the ATM format for presentation and use by the ACS and the destination IOC. A traffic policing capability is provided within each IOC, allowing the determination of conformance to a prespecified traffic profile.

Selective enabling of various policing modes is possible on a channel-specific basis, allowing each of the 4096 channels supported at an RXGA to be managed in a number of ways.

The policing strategies are part of a class of algorithms termed "leaky bucket" algorithms. These algorithms allow transmission as long as the source has credits. Credits are replenished at a predetermined rate, normally the average rate of the channel to be policed. However, only a finite number of credits is allowed to accumulate. Whenever a channel is checked for service and a unit of data is ready to be transmitted, credits are checked. If sufficient credits exist to satisfy the transmission, then it occurs. Otherwise the excess can be discarded or marked.

5.7.2   Cell Loss Priority Marking/Discrimination

For some traffic sources it is useful to have the capability to signify that some traffic is less important than other traffic, and should, therefore, be the first cells to be discarded in the event of congestion. Periodic background statistics reporting could be an example. The NCX 1E6 U-plane allows discrimination of traffic based on the input value of cell loss priority (CLP). In addition, the U-plane will have mechanisms to detect nonconforming traffic flows and mark them with the CLP bit, if required.

5.7.3   Traffic Shaping

In order to allow users to support different application types, there is a need to "shape" traffic in the U-plane. The intent is to allow operations from the purely statistical to the pseudo-deterministic. In a purely statistical operation, very high throughput and low delay traffic can be provided. However, the variance of these parameters around the mean is large. If a large population is combined, and the component sources approximate Poisson profiles, the typical load on the system is the sum of the means of the individual traffic components. Unfortunately, much of the access traffic to ATM networks originates on LANs, which don't typically exhibit Poisson behavior and generate large traffic volumes from a small number of currently active users.

In pseudo-deterministic operation, mean throughput for an individual channel is typically lower than can be achieved in pure statistical operation and delay is typically higher. However, the variance around the mean for these parameters is lower. This allows the statistics to be more accurately predicted than for an equivalent size purely statistical population. The traffic shaping capability of the NCX 1E6 allows for both types of users, which allows the user to define the expected traffic profile. When a channel is determined to be nonconforming to the defined profile, the RXGA discards the inbound traffic or marks it for discard in case of congestion. This marking information is conveyed to other system components to allow discrimination against nonconforming traffic flows in the presence of contention for system resources.

ATM networks based on an NCX 1E6 allow combinations of applications to be operated simultaneously. Some channels can be passed through the network in a purely statistical manner, others can be constrained to a pseudo-deterministic profile.

Traffic arriving at the shaper is physically buffered in memory until its appropriate transmit time. Since traffic is buffered, memory resources required for traffic shaping can exceed a preconfigured limit. In this case, newly arriving traffic is discarded.

As a result, SLAs that need a guaranteed level of service—and are willing to conform to an agreed throughput—have a higher probability that their traffic will achieve the requested service level. The downside is that there is very little potential for statistical gain of network resources in this traffic stream, since data are buffered and are not allowed to enter the network until permitted by the traffic shaper.

This compromise is between flow fluctuation and delay. SLAs can be designed to have high predictability, but the subscriber can be forced to tolerate a higher delay and a greater variation in that delay.

5.7.4 Discard/Marking Strategy

A key feature of the NCX 1E6 is its ability to determine which cells should be discarded—rather than queued—during a period of congestion. This decision process depends on several variables:

- Amount of traffic already buffered for a given output port
- Type of traffic to be queued
- Subscriber's QoS parameters

The NCX 1E6 takes special action for AAL5 frame-related traffic. When an exception occurs that causes the NCX 1E6 to discard a cell, subsequent cells will continue to be discarded until the end of frame (EOF) cell is detected, as described in Sec. 4.6.4.

The EOF cell is never discarded. The algorithm must be disabled for any single-segment traffic, i.e., AAL1. For AAL5, when the discard would start on the last cell of a frame, the mechanism will be internally defeated so that two frames are not lost when only one should have been. This mechanism ensures that once a cell that makes up part of a frame is discarded, other—now useless—cells also are discarded. At the same time, the EOF marker always is available so that frame boundaries are secure, and a complete frame is not lost simply because one cell was discarded. This intelligent cell discard implementation is one of the features contributing to the NCX 1E6's excellent congestion avoidance.

5.7.5 Cell Queuing

Each IOC contains up to 4 physical links, 1 CPU link, 1 multicast server link, and 20 discrete queues. Each queue can be logically associated with these link resources.

For a single physical link IOC, there typically will be four queues of 128 cells, eight queues of 512 cells, and four queues of 1024 cells. A group of one queue of 128 cells, two of 512, and one of 1024 are associated with a scan list and a meter, and are referred to as a Myrtle. This capability is shown in Fig. 5.13. Each queue is divided into discard priority zones, allowing queuing decisions to be made on traffic of differing service classes.

5.8 Node Commissioning

Each NCX 1E6 is delivered with its operating software preloaded and configured. Each NCX 1E6 is defined for a specific application and can be customized to meet an individual customer's specification. Each model will have been designed, sized, tested, and certified to provide a given level of performance and functionality in active service.

Figure 5.13

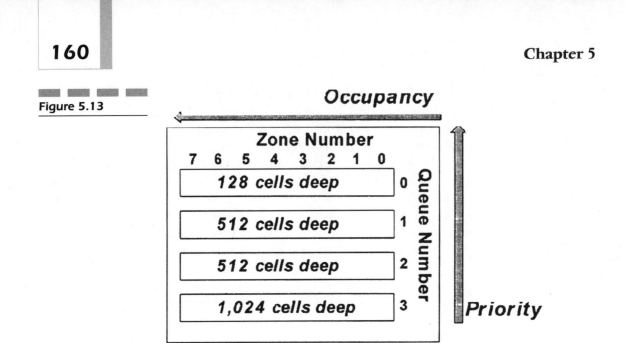

5.9 Configuration

NCX 1E6 nodes are initially characterized in the factory with a software image in flash EPROM. The factory software image fixes the software characterization for a specific protocol, and provides enough configuration information to support remote node initiation. On-site work is limited to physically locating, cabling, power-up, and establishment of deploy-time operating variables.

Telematics provides an extensive range of installation support services to ensure smooth and seamless network rollout. Node configuration consists of two steps: hardware and port configurations. The software image determines the configuration validation checks against installed hardware. Field-installed cards will be validated against the software image, with appropriate indications of success or incompatibilities to the operator.

OMS provides support in the preparation of node-specific configuration and in download. This will include node-specific software, in some cases.

OMS also provides support for inventory management, enabling a full report of the node hardware and software population to be generated that can be validated against order information. This facility also is valuable in the continuing management of the hardware and software resources that support the network.

5.10   NCX 1E6 Switch and Network Management

The increasing need for businesses and their customers/suppliers to exchange large volumes of data, along with the emergence of multimedia applications, is creating new requirements for broadband communication using ATM, SDH, and SONET technologies. With the growth and increasing complexity of new networks and the continued need for consolidation of legacy infrastructures, end-to-end network management is becoming critical for NSPs desiring to deliver services responsively and profitably.

To enable end-to-end provisioning, performance management, fault isolation, and problem resolution, Telematics is providing an OMS—a comprehensive, powerful network management software solution.

5.10.1   OMS Features

OMS simplifies the tasks of managing complex networks with intuitive graphical interfaces, cross-field data checking and correlation, context-sensitive selection, and advisory messages. OMS is designed with scalability and growth requirements built in from the start, so that NSP investment in managing networks—network management capital and personnel training—is preserved.

OMS consists of a set of applications and tools for the configuration, control, and monitoring of NCX 1E6 networks (see Fig. 5.14). It is based on HP OpenView and runs on Sun SPARCstation workstations.

Figure 5.14

The Fault Management System (FMS) module of OMS helps network administrators identify and locate specific system and network failures in the shortest possible time with the greatest possible accuracy.

The Change Management System (CMS) module of OMS enables the network operator to:

■ View and modify network elements, including user and service attributes.

■ Manage the steps involved in specifying, preparing, implementing, and reporting network changes, including rollback capabilities.

5.10.2   HP OpenView

Built on the popular HP OpenView management framework, OMS readily integrates with enterprisewide management or telecommunications management networks. The SNMP management protocol, LAN-based File Transfer (FTP), SQL data access, and X-Windows/Motif style graphical presentation (X11R5, Motif 1.2) all are supported by OMS. Together they provide the building blocks to facilitate integration of OMS into the existing management infrastructure of an NSP.

HP OpenView's graphical user interface also provides sophisticated capabilities for quick and simple task execution.

HP OpenView Network Node Manager provides fault, configuration, and performance management, acting as the OMS framework. Network Node Manager includes the following features:

■ Automatic node discovery (IP-addressable nodes)

■ MIB loader and browser

■ Event manager

■ End-user customization

■ Map creation with real-time updating

■ Application builder—single-view management accomplished by integrating existing and third-party or custom applications in a matter of minutes without programming

Figure 5.15 provides a logical view of the HP OpenView platform, Telematics' applications, and their logical connection to an NCX 1E6.

FMS is specifically designed to manage Telematics' equipment. Additional features are provided by integrating this application with the graphical interface and software architecture provided by HP

Figure 5.15

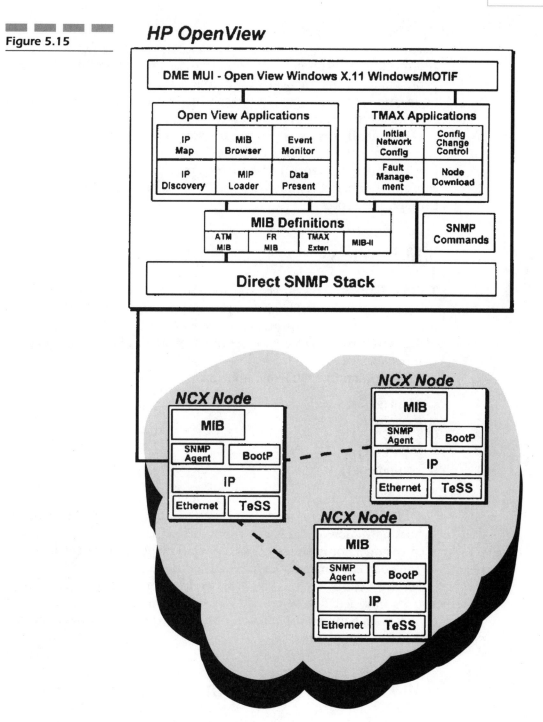

OpenView. FMS supports a complete graphical network topology representation, allowing operators to efficiently monitor network status and quickly focus on network trouble areas. Overall network topology is presented in hierarchical views, from a top-level map to individual node- and card-level views. Network maps can be customized to match customer network configuration.

OMS initially discovers existing network topology, and then continually monitors the network for changes in node status and network topology, updating network maps in response to network events and a configurable periodic poll. All NCX 1E6s automatically are discovered through this process.

5.10.3 Views

This application allows users to display and modify the network maps using a graphical interface. Network topology is presented in a series of hierarchical views, beginning with a geographical map. Each view is displayed as a separate window. If required, views can be organized as virtual network views.

Views contain the following information:

- Maps (global, regional)
- Sites
- Links
- Equipment
- Services

OMS allows operators to navigate through network maps to get increasingly more detail about the nodes. At the Telematics node-level map, the operator is presented with a window providing a detailed view of the node equipment, node services, and node events.

Service objects, such as software services, frame relay, or ATM UNI, also are modeled, discovered, and maintained in the database. All logical connections between the physical nodes and the service objects are represented as logical links.

5.10.4 Event Monitoring

Events are reported to the network management station as they occur in the network. Events include any operator-initiated action, network-generated fault messages, and alarms—traps—that are event subsets. Traps can be associated with physical or logical objects.

Events are reported to the management system as traps. This enables accurate status display of affected elements on graphical maps, which can then be logged for historic reporting. Various filtering and correlation techniques can be applied to the events to reduce the amount of information presented to the operator or to transform the events into actions.

Once generated, traps from the network objects can be processed in a variety of ways, and the display customized to show

- All events
- Filtered events
- Equipment and location
- Severity, shown by color code
- Programmable severity levels

A number of event types are recognized by the application, including

- Failure of transmission resources
- Loss of synchronization
- Unmet QoS parameters

5.10.5   Performance Management

Performance management contains tools that support the creation of functions for retrieval and display of statistics and performance data.

The performance management tools allow the users to collect and graph statistical data in real time, and to build diverse applications to extend the capabilities of the management environment. The ability to trap on exceeded threshold can be built through the application's template feature. Collected data are stored in a file and can be exported to external applications for performance trending analysis and other statistical operations. All statistics parameters defined as MIB objects can be selected for retrieval, displayed on demand, or stored for later review and analysis. Standard MIB variables are employed where possible to provide for common metrics across protocols and equipment type.

The performance management features include

- Real-time statistics gathering and viewing
- Real-time graphical displays

- Graphical files archiving
- Scheduled statistics collection to a file or a display
- HP or independent vendor software applications
- Customer-developed applications
- File export to trending or other performance management applications such as capacity planning and modeling

OAM Flows OAM support includes detection and processing of the ATM F4 and F5 management flows, where the information arrives in-band and is indicated by either the VCI=3 or VCI=4 for F4 flows or by the 3-bit PTI field code points 4 and 5 for F5 flows.

Where the termination of an OAM flow within an NCX 1E6 is desired, and a valid destination routing entry is discovered, OAM cells are processed according to ITU-T I.610.

CMS is the NCX 1E6's Configuration Management and Change Control System, encompassing inventory and configuration management capabilities by tracking and controlling each physical (node, card, port) and logical (user, service, application) asset on the network.

The configuration database resides in the Ingres database, which supports SQL requests for network configuration data or report generation. User privileges are validated to prevent unauthorized entries or changes to the configuration database. SQL data requests can be used to create inventory checks of network equipment or to export network configurations to external network design tools to support future network expansion planning processes.

OMS also provides a scripting interface that contains commands that allow configuration data to be created, modified, or deleted from the CMS database.

Relational database management systems (RDBMSs) extend the scope of configuration and inventory management by automatically updating corresponding system modules when changes are made. The RDBMS also links graphic and text databases, giving a user a visual or text record of the network configuration.

5.10.6 Configuration Management

OMS includes applications for configuration management and change control designed to reduce the operating costs of service networks. This is scheduled to be provided in two phases. This section describes the complete functionality that will be delivered.

OMS provides a structured model of entities that make up a network's configuration, along with a complete set of graphic screens to create and update the network configuration entities. The screens provide access to node-, card-, and device-level entities.

CMS is based on an Ingres database accessed using SQL commands.

CMS allows configuration changes to be made prior to an implementation date. The change implementation can then be carried out automatically by the operator at a predefined time.

5.10.7 Change Control

The primary function of the Telematics Configuration Management Application is managing configuration changes.

The configuration management process is closely connected to the change management process. At update delivery time, the set of updated CMS entities is automatically applied to the network through the following steps:

- Check the consistency of the requested entity changes.
- Generate new configuration files from the database description of the entity.
- Download these files to the SNMP agent. These files then will be appropriately handled by the agent to deliver the changes or the entire configuration.
- Issue commands to cause real-time changes in the nodes.
- If the change is successfully activated, update the CMS database to reflect that the change has actually been made in the network.

The process begins with a change request (CR), which specifies the desired changes to individual network elements and groups related changes for batch application. The application manages the CR and the introduction of the configuration changes into the network, providing administrators with tools to

- Provide a standard mechanism (CR) for defining, updating, delivering, and activating configuration changes to the network.
- Ensure that changes are not carried out until they are approved by an authorized agent.

- Track a number of individual updates within a single CR.
- View the status of the CR to determine if the change is pending, successfully completed, or failed.

5.10.8 Statistics

Statistics are available for all resources through MIBs defined in the SNMP agent and accessed from the OMS applications. Statistical values enable the application to quantify and measure QoS parameters against SLAs, as well as determine network resource utilization. A comprehensive set includes

- Low-level statistics
- Calculated variables
- Node resource statistics
- Link/trunk resource statistics

Because OMS can normalize access across all network equipment, these applications further allow the users to consolidate operations on fewer platforms with fewer interfaces. This will allow the staff to focus on other areas of network planning and management.

The advantage of standards and open interfaces can be summarized as

- Improved personnel productivity and ease of providing new services with improved network configuration tools.
- Integration and management of multivendor networks.
- Availability of applications from the independent software vendors (ISV).
- Merging of standards-based and legacy networks. Users can preserve their investment in legacy systems and network management applications while migrating to industry-standard open network management.

6

Case Study: Frame Relay–ATM Internetworking— Ascend Communications B-STDX 8000/9000

This chapter is presented as an example of a real-world application of frame relay—ATM interworking technology in a service provider context.

6.1 About Ascend Communications Corporation

Ascend Communications, Inc., is at the forefront of broadband data communications with its family of scalable, carrier-class frame relay, ATM, IP switching, and remote access products. Used by public carriers and Internet service providers worldwide, Ascend products are forming the foundation of the public network infrastructure for global communications. Ascend is headquartered in Alameda, California, with offices throughout the Americas, Europe, and Asia/Pacific. For more information about Ascend and its products, please visit the Ascend Web site at http://www.ascend.com.

6.2 B-STDX 8000/9000 Platform

The B-STDX is a broadband packet switch engineered to support true multiservice networking. The B-STDX products provide a scalable, flexible, and cost-effective multiservice WAN platform for the delivery of frame relay, ATM data exchange interface/frame UNI (DXI/FUNI), and SMDS DXI services.

The B-STDX hardware platform is based on a 1.2-Gbps midplane bus. The product family provides fully redundant, high-performance capabilities, utilizing standards-based foundations. Two models are available, which differ primarily in terms of the number of input/output (I/O) slots supported: the Model 9000 and the Model 8000.

The B-STDX 9000 is a 16-slot, high-capacity, modular multiservice platform packaged to accommodate large networks with high-density requirements.

The B-STDX 8000 is an 8-slot, modular multiservice platform providing the right balance of price and performance. The 8000 is designed to provide a cost-effective solution to meet the rapidly growing needs of

Many thanks to Jim Mathison and Jim Martel (both of Ascend Communications, Inc.) for their significant contributions to this chapter low- to medium-density sites within public network carriers and enterprise end-user networks.

6.3   B-STDX Hardware Architecture

6.3.1   Principal Components

The B-STDX hardware architecture has three primary components (see Fig. 6.1):

- *Control processor (CP).*   Provides background management and static networking functions in support of the real-time networking functionality provided by multiple IOPs.

- *I/O processor (IOP).*   Manages the lowest level of a node's trunk or user interfaces. It performs physical data link (frame/cell) and multiplexing

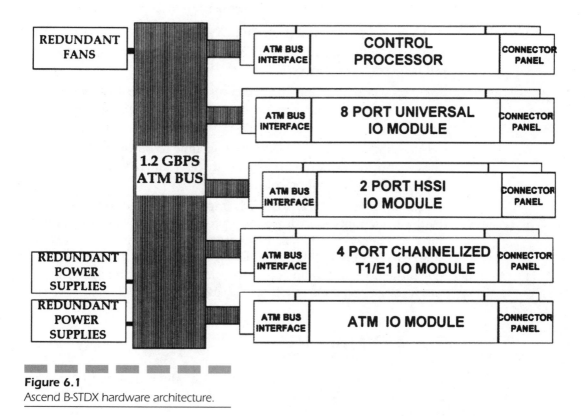

Figure 6.1
Ascend B-STDX hardware architecture.

operations on external trunks and user links. I/O adapter modules (IOAs) are used to connect the various IOPs to the network.

■ *System enclosure.*   One common enclosure and power/cooling system with optional power redundancy.

The hardware architecture employs symmetrical reduced instruction set computing (RISC) multiprocessing that consists of a CP interacting with multiple I/O modules.

The CP and I/O modules utilize the Intel i960 RISC processor to deliver scalable, high-performance packet switching needed for ATM and high-bandwidth transmission environments, such as HSSI, T3/E3, OC3c/STM-1, or high-port-density configurations.

6.3.2   Network Resiliency Features

Because Ascend switches are designed to serve as a platform for the delivery of robust public data services, the B-STDX is engineered to provide a high degree of reliability and networkwide availability. Resiliency features include

■ *Redundancy.*   Fully redundant CP, I/O modules, power supplies, and fan modules for high reliability in mission-critical applications. All failovers are automatic.

■ *Hot swap.*   Live insertion and hot swap of all CP and I/O modules, power supplies, and fan modules to ensure maximum network availability.

■ *Sophisticated, resilient circuit routing.*   Dynamic rerouting of PVCs in the event of network outage using fully distributed Virtual Network Navigator circuit routing technology.

■ *Midplane architecture.*   The B-STDX implements a midplane hardware architecture that separates processor modules from I/O interfaces, allowing any failed I/O processor, I/O adapter, or power supply to be removed (hot swapped) without having to power down the unit or disconnect any of the cables, resulting in a low mean time to repair (MTTR) (see Fig. 6.2).

6.3.3   I/O Modules

The B-STDX accommodates up to 14 I/O modules that can be configured noncontiguously in single or redundant configurations. Redun-

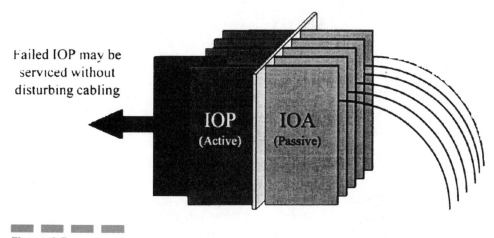

Failed IOP may be serviced without disturbing cabling

Figure 6.2
Midplane architecture.

dant configurations consume two neighboring I/O processor slots and one combined I/O adapter slot. A variety of I/O modules accommodate numerous interface specifications, speeds, and protocols. Both the control processor and the I/O modules utilize passive I/O adapters that allow any failed I/O module, power supply, or fan module to be removed under power without disconnecting cables. This results in a low MTTR for the B-STDX system.

The B-STDX supports the following types of I/O modules:

Frame Services

- 2-port HSSI module
- 4-port channelized T1 module
- 12-port E1 module
- 4-port T1 ISDN PRI module
- 4-port E1 ISDN PRI module
- 8-port universal (V.35 or X.21) module
- 10-port DSX-1 module
- 1-port channelized DS3 module

ATM Services

- 1-port ATM CS DS3/E3 interworking module
- 1-port ATM IWU OC3c/STM1 interworking module

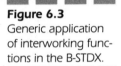

Figure 6.3
Generic application
of interworking func-
tions in the B-STDX.

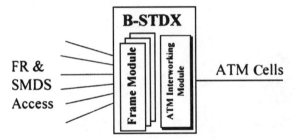

The B-STDX performs frame relay–ATM interworking functions on
the ATM CS and ATM IWU modules. Traffic originating from frame-
based modules is aggregated and passed to the ATM interworking card,
which performs FRF.5 or FRF.8 interworking functions and supports an
ATM physical interface (see Fig. 6.3).

6.3.4 Data Flow through the B-STDX

This section describes the flow of data through an Ascend frame
relay–ATM network. Specifically, it discusses how frames are received
from a frame relay DTE over a data link connection identifier (DLCI)
at a B-STDX, processed by a B-STDX switch, relayed over a trunk connection to another B-STDX, processed by a second switch, and transferred to a destination frame relay DTE over another DLCI. In addition
to frame relay packet flow, the differences in this packet flow for
frame relay DTE–to–frame relay network-to-network interface (NNI),
frame relay–to–ATM, and ATM DTE–to–ATM DTE data flows are
also addressed.

B-STDX Packet Flow   A permanent virtual circuit (PVC) is configured
by a network administrator linking a DLCI established at one location
in the network to a second DLCI established at another location in the
network. Virtual Network Navigator [Ascend's Open Shortest Path First
(OSPF)–based dynamic distributed routing technology; see Sec. 6.4] determines the best path through the network regardless of the number of
hops, and a virtual circuit is then established. A virtual channel number,
including a priority level (1 to 4), is then assigned. The B-STDX establishes a configuration table that keeps track of the association between each
DLCI and its associated location within the switch, by port number and
chassis slot number. This information is stored in nonvolatile RAM,

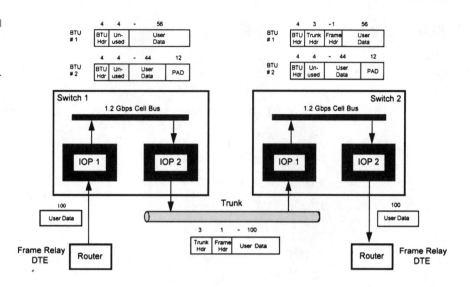

Figure 6.4
Data flow through
an Ascend network.

known as PRAM. Once a PVC is established, packets may begin flowing over the B-STDX network using the selected route.

The B-STDX is based on an output buffering switch architecture. Because of the speed of the processing on the I/O processor and the bandwidth of the bus, only temporary buffers long enough to hold a single bus transfer unit (BTU) of 64 bytes are required during receipt of a frame from a frame relay DTE. Frame relay packets are received at an IOP in 56-byte increments. An 8-byte header is added to the incoming data, forming a fixed-length 64-byte BTU. As shown in Fig. 6.4, padding is used to fill the last BTU of the frame to maintain constant 64-byte BTUs through the B-STDX. BTUs are then switched over the B-STDX bus to the destination IOP based on the VC number and slot number listed in the configuration table. The 1.2-Gbps B-STDX bus is capable of supporting approximately 2.3 million BTU transfers per second or about 0.8 million 128-byte packets of frame relay data per second.

Table 6.1 shows the header format for the 4-byte (32-bit) BTU header. An additional 4 bytes of header is added and reserved for future use. Bits 7 and 8 are used for frame relay prioritization, allowing up to four priority levels to be assigned to the frame relay connection. (Priority 1 is always used for management information.) The priority level is established at connection setup time. Bits 9 through 21 (13 bits) are used for VC assignment, allowing the IOP architecturally to support up to 8192 DLCIs (connections from DTEs to the local IOP). Up to 1024 DLCIs (the ANSI frame relay standard maximum) are supported per port. Bits 22

TABLE 6.1

BTU Header
Format

Bits	Interpretation
0	SOM (start of message)
1	EOM (end of message)
2	CTL [control (nonuser data)]
3–6	Sequence number
7–8	FR: priority; SMDS: in link type
9–21	FR: VC ID; SMDS: in link index
22	ATM cell indicator
23	SMDS
24–29	Data length
30	FR: start of QuickPath segment
31	FR: BECN bit

and 23 are used to indicate ATM cells and SMDS DXI frames, respectively, changing the use of bits 7 through 21 when the switch is used for ATM or SMDS traffic.

BTUs are chained together in the outgoing IOP until a full frame is received from the DTE. Within the IOP buffer pool, traffic is mapped into one of four queues based on the assigned priority. The BTU header is then removed, and a trunk header and frame header are added. Under normal operation, all frame relay frames transferred over trunk connections contain both trunk and frame headers. When the B-STDX's Quick-Path fragmentation feature is enabled (in order to minimize latency through the network), the frame header is used only during the first segment of a frame relay frame. Following segments require only the trunk header.

Tables 6.2 and 6.3 show the header format for the trunk and frame headers. The VC ID is 15 bits long, architecturally supporting up to 32,766 virtual channels over a single trunk.

The SOM, EOM, and sequence ID fields are used to support Quick-Path operation. The red, amber, and green (when red and amber fields are set to 0) fields in the frame header are an expansion of the single-bit Discard Eligibility (DE) field present in the frame relay header received from the frame relay DTE at ingress to the network. In addition to the forward and backward explicit congestion notification (FECN and

TABLE 6.2

Trunk Header

Bits	Interpretation
0	CTL [control message (nonuser data)]
1	SOM (start of message)
2	EOM (end of message)
3–6	4-bit sequence ID
7–8	2-bit priority
9–23	VC ID

TABLE 6.3

Frame Header

Bits	Interpretation
0	CTL [control message (nonuser data)]
1	C/R (C/R bit from user's DLCI)
2	ODE (red frame)
3–6	DE (amber frame)
7–8	BECN
9–23	FECN

BECN) fields, these fields are important parts of the Ascend congestion management mechanism.

After removing the BTU header and adding the trunk and frame headers, the IOP services the queues in order of priority, transmitting frames out the correct I/O port as defined in the configuration table for the virtual channel.

Once the frame reaches the destination B-STDX over the trunk connection, the packet segmentation process is repeated and a BTU header is appended (see Fig. 6.4). The BTU(s) are then switched over the B-STDX bus until they reach the destination IOP, where they are stored in a common buffer pool. The BTU header, trunk header, and frame header are then removed, the frame relay header is added, and the frame relay frame is relayed to the destination frame relay DTE out the appropriate port and DLCI.

Frame Relay DTE-to-Frame Relay NNI Data Flow The flow of frame relay data from a frame relay DTE to a frame relay network-to-network interface (NNI) through an Ascend B-STDX network is identical to that described above for frame relay DTE–to–frame relay DTE connections. The differences are in the exchange of management information between a frame relay DCE (B-STDX switch) and the frame relay DTE (e.g., router) as compared to that between a frame relay DCE and another frame relay DCE (a network-to-network connection). For a discussion of this information exchange, refer to ANSI T1.617 Annex D.

Frame Relay–to–ATM Data Flow   There are three types of frame relay–to–ATM data paths supported by an Ascend B-STDX network:

1. The connection of a frame relay DTE to an ATM DTE

2. Frame relay–ATM service interworking between a frame relay DCE (a B-STDX switch) and an ATM DTE (e.g., host) over an ATM network

3. Connection of a frame relay DCE (B-STDX switch) to a frame relay DCE (another B-STDX switch) over an ATM network where the ATM network is used as a trunk connection between B-STDX switches

In connection type 1, traffic flow is similar to that in the frame relay DTE–to–frame relay DTE connection, discussed earlier. The modifications occur at the edge of the network. During virtual channel setup time, a frame relay DLCI is mapped to an ATM virtual path and virtual channel identifier VPI/VCI. At the ends of the network the appropriate frame relay header, including DLCI number, or ATM header, including VPI/VCI number, is inserted into the data exiting the network. Frame relay frames are segmented into ATM AAL5 (ATM Adaptation Layer type 5) cells at the ATM interface. Also, ATM cells are assembled into frame relay frames at the ATM interface. That is, the segmentation and reassembly process, known as the SAR function, is performed at the ATM interface for frame relay–to–ATM and ATM-to–frame relay communications. The B-STDX provides both IETF RFC-1483 to RFC-1490 translation and transparent support of RFC-1490 frames over an ATM connection.

In the second type of frame relay–to–ATM communications, where the ATM host is not attached directly to the B-STDX switch but rather over an ATM network, the process is similar to that for connection type 1 above. The only differences arise in that a VPI/VCI is established between the B-STDX and an ATM switch rather than with the ATM host directly.

In the third type of frame relay–to–ATM communications, two B-STDX switches provide communications for two frame relay DTE devices over an intermediate ATM network. This support is made possible by a B-STDX feature known as ATM OPTimum (open packet trunking) frame trunk, which allows B-STDX switches to use an intervening network for the purpose of interswitch trunking. In this case, the data flow is similar to that with other trunk connections (B-STDX to B-STDX) except that the entire trunk packet is segmented into one or more 48-byte ATM cell payloads, an ATM header is added, and cells are transferred over an ATM VPI/VCI established between adjacent B-STDX switches over an intervening ATM network. At the other end of the ATM connection, the ATM header is removed, the trunk packet is reassembled, a BTU header is appended to the trunk header, and the frame relay frame is switched over the B-STDX bus to the destination IOP. (See Sec. 6.7 for additional discussion on the subject of OPTimum trunking.)

ATM-to-ATM Data Flow   ATM-to-ATM data flow through the B-STDX switch, more commonly known as cell switching, is similar to frame relay switching through the switch, but with some important differences. Since all cells received from a DTE are already smaller than the data portion of the BTU frame, no segmentation by the IOP upon receipt of an ATM cell is necessary. The 53-byte ATM cell is inserted into the 56-byte data portion with 3 additional bytes of padding, a BTU header is added, and the cell is switched over the B-STDX bus to the destination IOP.

Cell-switched traffic is also not mixed with frame-switched traffic. For example, a DS3 ATM module may be used to concentrate cells to an OC3c ATM module (a cell-switching application). Likewise, a channelized T1 module can be used to concentrate frame relay frames to the same OC3c ATM module (a FR-ATM service interworking application). However, cells from the DS3 ATM module will not be serviced in the same fashion (e.g., queue prioritization) as frames that originate from the channelized T1 module. Each of the traffic sources will be handled separately on the outbound IOP in order to meet ATM's more strict QoS requirements. The Ascend B-STDX fully supports ATM QoS definitions. At connection setup time, the correct QoS is established through the network for each VPI/VCI.

6.4   Virtual Network Navigator

6.4.1   Overview

Virtual Network Navigator (VNN) is a sophisticated virtual circuit routing technology. Based on the industry standard OSPF protocol, VNN was first developed to support frame relay and IP services and has since been extended to support the rigorous QoS requirements of ATM. Like OSPF, VNN is a *link-state* routing technology. Link-state routing protocols are characterized by their excellent fault tolerance, rapid network convergence performance, and inherent scalability.

VNN also provides for efficient management of precious wide area bandwidth, through support of the following functions:

- *Automatic rerouting.* VNN will reestablish VCs over alternative paths in the event of a facility failure.
- *Policy-based routing*
  - *Cost-based routing.* VNN allows for definition of an administrative cost metric on each trunk for cost-based routing.
  - *Virtual private networks.* VNN provides these via designation of customer- or application-specific trunks.
  - *Management-only trunks.* VNN offers a set of trunks that are restricted to carrying management traffic only.
- *Load balancing.* VNN provides for an effective means for balancing traffic across parallel trunks.

Integration with ATM's call admission control function assures that the QoS requirements of each VC will be guaranteed from end to end, and that VCs that cannot be supported because of insufficient resources will be prevented.

6.4.2 Functional Description

VNN has three principal components:

1. *Exchange of "hellos."* Adjacent switches exchange keep-alive PDUs.
2. *Advertisement of local topology.* Switches broadcast *link-state advertisements* (LSAs) to all switches in the domain. LSAs include link-specific metrics for available bandwidth, delay, and loss on a QoS class basis. Switches receive and aggregate data from LSAs to build a complete network topology database. A master copy of the topology database resides on the CP module. Copies of the database are then distributed to each line card for scalable VC setup performance.
3. *Best route calculation.* Each line card in the B-STDX or CBX performs "best path" calculations independently using a Dijkstra algorithm.

Once a best path is selected, a VC is established automatically one hop at a time across the designated path. As the call setup proceeds, each node along the path verifies that the resources are indeed available before the setup proceeds to the next hop. In the event that sufficient resources are not found, the network routing topology will be updated via an LSA. A new path from source to destination is then calculated using the updated routing information.

6.4.3   Relevance of VNN in a Frame Relay–ATM Context

In a mixed frame relay and ATM network, VNN runs seamlessly across all nodes, allowing for simple one-step provision of virtual circuits by selection of the circuit endpoints via the Network Management System (NMS) graphical user interface (GUI). VNN automatically invokes frame relay–ATM interworking functions across the designated path, including transparently mapping the frame relay traffic contract [expressed in terms of committed information rate (CIR), excess burst capacity (Be), and committed burst capacity (Bc)] to the equivalent ATM traffic contract [expressed in terms of peak cell rate (PCR), sustainable cell rate (SCR), and maximum burst size (MBS)] for bandwidth reservation purposes. This eliminates the need for separate configuration of frame and ATM segments, speeding service delivery and reducing the overall cost of operations. These characteristics make VNN a strong tool for competitive advantage for public data service providers.

6.5   Interworking Product Overview

The ATM CS and ATM IWU cards represent Ascend's second generation of ATM technology for the B-STDX. These cards build upon Ascend interworking technology by adding support for pure cell switching to the B-STDX platform.

The ATM CS and ATM IWU cards share the same fundamental internal architecture, which supports the following features:

■ Frame relay/ATM network interworking (per FRF.5)

■ Frame relay/ATM service interworking (per FRF.8)
■ ATM OPTimum frame trunking
■ ATM OPTimum cell trunking
■ Direct cell trunking capability for seamless integration between the B-STDX and the CBX 500 ATM switch

6.5.1 ATM CS Module

The ATM CS (cell switching) module is a one-port DS3/E3 ATM card. This card supports both ATM UNI DTE and ATM UNI DCE logical port types and is available in both single and redundant configurations.

The hardware design of the CS and IWU cards provides for isolation of ATM traffic from frame relay interworking traffic through separate data paths. A high-level architectural diagram of the card is shown in Fig. 6.5.

6.5.2 ATM IWU Module

The ATM Interworking Unit (IWU) supports a single optical-fiber (OC3c/STM1) physical port at speeds up to 155 Mbps. A common IOM supports both OC3c and STM1, selected via software during installation and configuration. Both single-mode and multimode fiber versions are offered. The optical-fiber interface is rated for medium-reach applica-

Figure 6.5
Architecture of ATM CS/IWU modules.

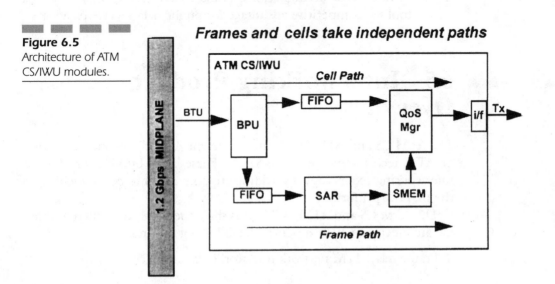

tions (up to 20 km). The card supports both single and redundant configurations, and may act as either an access or a trunk interface. The ATM IWU is most frequently used for high-speed trunk connectivity to the CBX 500 ATM switch, or for high-speed connectivity to a router in IP networks.

The redundant version of the ATM IWU offers a unique hardware design that occupies two chassis slots. Consistent with earlier generation IO modules from Ascend, the ATM IWU redundant assembly protects against IOP failure (see Fig. 6.6).

The redundant ATM IWU also provides protection against failure of the optical transceiver components on the I/O adapter that support the single or multimode fiber interface (see Fig. 6.7). Each redundant ATM IWU I/O adapter assembly supports two optical transceiver units.

In the event of a transceiver failure, the failed unit can be replaced in the field *while the module remains in operation,* providing a significant advantage in service provider and mission-critical environments where nonstop operation is critical.

Figure 6.6
ATM CS/IWU redundant module design.

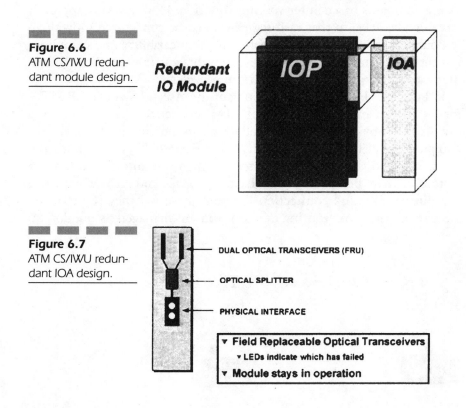

Figure 6.7
ATM CS/IWU redundant IOA design.

6.6 Applications

6.6.1 Trunking

Trunks connecting Ascend switches are used as network resources that carry traffic from multiple end-user sources for intra/internetwork communication between the switches and for automatic rerouting of data in the event of network outages.

Ascend cell trunks allow the customer to quickly add switches to the network. Once a switch is installed and added to the NMS map, and a trunk configured to the existing network, Ascend's OSPF-based Virtual Network Navigator immediately propagates this information network-wide for the purpose of connection routing.

When the interface between the FR network and the ATM network is an ATM UNI port, service interworking PVCs must be implemented as two separate PVCs, one in the FR network and one in the ATM network (see Fig. 6.8). This approach has the advantage of allowing the FR and ATM equipment to be manufactured by different vendors but still provide a standards-based interworking function. However, this approach has a disadvantage in that each segment must be configured individually, doubling the configuration effort, reducing scalability, and complicating the circuit maintenance. These disadvantages translate directly into increased operations cost for the service provider.

If both the FR and ATM equipment are Ascend switches (B-STDX for FR service and the CBX-500 for ATM service), service interworking can also be provided by a direct cell trunk connection between the FR equipment and the ATM equipment (see Fig. 6.9). Note that both the FR switch and the ATM switch are shown in one integrated network instead of there being two separate clouds for FR and ATM. Since there is a direct cell trunk connection between these switches, the switches are part of the same routing domain, and are managed as one logical

Figure 6.8
Service interworking across an ATM UNI interface.

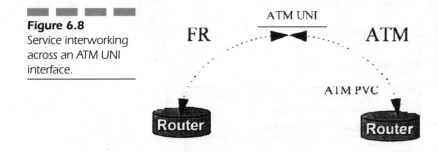

Figure 6.9
Service interworking
across an Ascend
direct cell trunk.

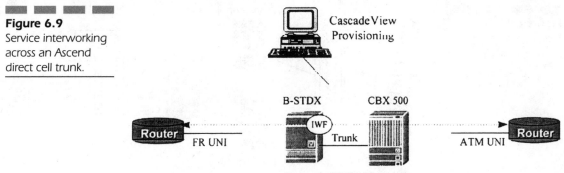

Cascade VNN Network

network even though multiple services are supported. With this configuration, the user needs to configure only one PVC to establish a service interworking circuit. This *multiservice PVC* is FR on one end and ATM on the other end, and the network will transparently and automatically manage the transition between the two technologies.

The use of a cell trunk to support FR-ATM service interworking provides the following benefits to the service provider:

1. Service interworking PVCs are configured in one step via the NMS GUI. The user need only select the two endpoints. The NMS will sense the technology at the endpoints and request the necessary information. Thus, FR, ATM, and FR-ATM service interworking PVCs are configured in an identical manner, reducing training costs and simplifying the configuration process.

2. Multiservice PVCs will reduce the total number of circuits that must be provided and thus will decrease operations cost and increase network scalability.

3. The multiservice PVC is managed from one NMS platform, and will present the user with one status for the integrated PVC. The user will not need to correlate status from two separate PVCs to determine the health of the end-to-end connection.

Direct Cell Trunk   B-STDX Release 4.2 adds support for Ascend ATM cell trunking capability to the B-STDX platform. Cell trunking may be used to provide for deterministic performance between two B-STDX switches or between B-STDX and the CBX 500.

Use of the direct cell trunk (see Fig. 6.10) achieves a seamless integration of the B-STDX and the CBX 500 in a single network. Ascend's OSPF-based Virtual Network Navigator sees this integrated network as a single

Figure 6.10
Direct cell trunking
between B-STDX and
CBX 500.

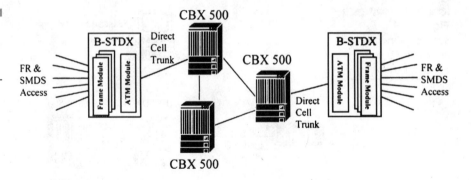

routing domain, under which all Ascend switches function as peers. This eliminates the need for ATM OPTimum trunking, and allows for faster provision and ease of maintenance.

Direct cell trunk fully supports the establishment of permanent virtual circuits between two B-STDX access ports across a CBX 500 core, as well as circuits that originate on a B-STDX and terminate on the CBX 500.

Although ATM cell trunks between B-STDX switches are supported (see Fig. 6.11), customers are encouraged to select the CBX 500 for ATM backbone switching rather than attempting to build a large-scale ATM switching infrastructure using the B-STDX alone. The CBX 500 brings a superior QoS feature set to provide a guaranteed level of service across all classes of ATM traffic.

ATM OPTimum Trunk *OPTimum (open packet trunking)* is a unique software component that allows public data networks based on frame relay or ATM to be used as trunk connections between Ascend switches.

ATM OPTimum trunks (see Fig. 6.12) provide a cost-effective means of passing Ascend logical trunks across a third-party ATM network infrastructure. This approach has clear attractions when leased-line costs for trunking are high, and is often the most practical means for networking B-STDX switches across another vendor's ATM network.

Figure 6.11
Direct cell trunking
between B-STDX
switches.

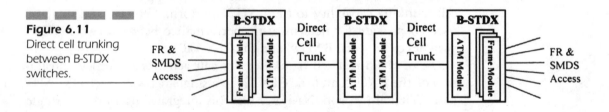

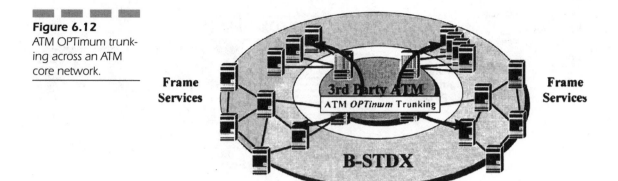

Figure 6.12
ATM OPTimum trunking across an ATM core network.

The ATM CS and IWU cards support two different logical port (Lport) versions of the ATM OPTimum trunk:

- ATM OPTimum frame trunk: using a single ATM virtual channel connection (VCC tunneling)
- ATM OPTimum cell trunk: using a single ATM virtual path connection (VPC tunneling)

The B-STDX ATM OPTimum frame trunk Lport type multiplexes a number of frame relay virtual circuits onto a single ATM virtual circuit connection (see Fig. 6.13). The Ascend trunk protocol header is encapsulated in an ATM cell for transmission on the physical ATM UNI interface.

The ATM OPTimum cell trunk multiplexes multiple frame relay virtual connections into a single ATM VPC for tunneling across an intervening ATM network (see Fig. 6.14). Each frame relay VC is mapped to a corresponding ATM virtual channel connection within the virtual path. The VCI assignment is made sequentially by the switch software and does not require operator intervention.

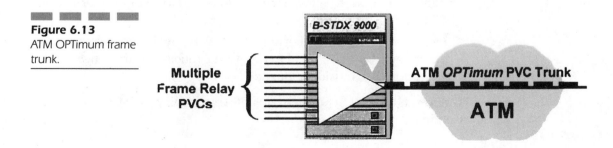

Figure 6.13
ATM OPTimum frame trunk.

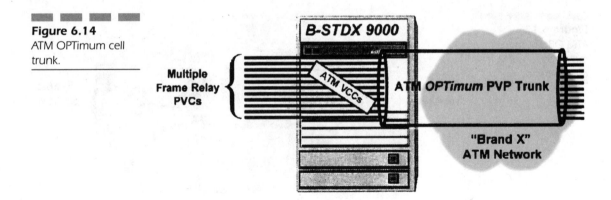

Figure 6.14
ATM OPTimum cell trunk.

6.6.2 Service Interworking

Ascend takes justifiable pride in its strong heritage of support for frame relay/ATM service interworking. Since late 1994, Ascend has demonstrated its strength in deploying carrier-class interworking solutions that allow subscribers to make a gradual migration to ATM, while maintaining cost-efficient connectivity to the bulk of their remote sites using frame relay.

In support for interworking, the following basic functions are performed:

- *Circuit traffic parameter mapping.* Frame relay traffic parameters (CIR, Bc, and Be) are mapped to the equivalent ATM parameters (PCR, SCR, and MBS).

- *Upper-layer protocol translation.* RFC-1490 to RFC-1483 translation.

- *Optional traffic shaping.* This assures that circuit behavior will not violate the behavior expected from the ATM network as expressed in the ATM traffic contract for the VPI/VCI.

- *DE-to-CLP mapping.* Forward and reverse dynamic mapping of the frame relay Discard Eligible (DE) bit to the ATM cell loss priority (CLP) bit are provided.

- *FECN-to-EFCI mapping.* Forward and reverse dynamic mapping of the frame relay FECN field to the ATM explicit forward congestion indication (EFCI) field are provided.

Ascend supports two scenarios of frame relay–ATM service interworking:

1. *FR-ATM service interworking over an ATM UNI.* This is used when interfacing an Ascend frame relay network to a third-party ATM network (see Fig. 6.15). In this scenario, separate provision of the ATM segment and the frame relay segment is required, as there are two networks, each under its own administrative control.

2. *FR-ATM service interworking over an Ascend trunk* (see Fig. 6.16). This is used when connecting a frame relay UNI to an ATM UNI in an all-Ascend network, over either a direct cell trunk or ATM OPTimum cell trunk. In this scenario, service interworking PVCs are provided by defining endpoints for the virtual circuit. One endpoint is provided as a frame relay Lport, and the opposite end of the circuit is defined as an ATM UNI. The Ascend network will recognize the circuit as an interworking circuit, invoke the Ascend frame relay–ATM service interworking functionality automatically, and determine the best path through the Ascend network with full Virtual Network Navigator functionality.

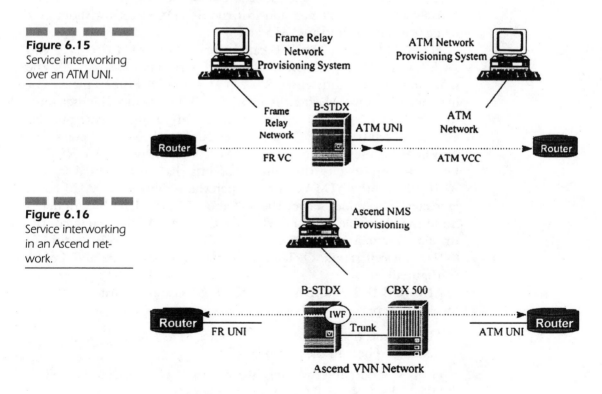

Figure 6.15
Service interworking over an ATM UNI.

Figure 6.16
Service interworking in an Ascend network.

6.7 Logical Port Types

Ascend switch products are designed with a concept of a *logical port,* or Lport. This architecture provides flexibility to the user by allowing different software images and interface definitions to be applied to a given *physical port* (Pport). In the case of the ATM CS and ATM IWU cards, as many as 124 individual Lports may be created on a given Pport, although in most applications a single Lport image is used.

6.7.1 ATM UNI DTE/DCE

The previous generation of ATM cards for the B-STDX provided support only for FR-ATM interworking, and thus were only capable of interpreting the AAL5 Adaptation Layer. The ATM CS and ATM IWU modules will perform FR-ATM interworking, but also provide support for native cell switching. With these cards, the switch can support any Adaptation Layer, not just AAL5. Cells can come into the B-STDX, be switched natively across the cell bus, and exit the B-STDX as cells without ever being reassembled into an AAL5 PDU.

The ATM CS and ATM IWU cards support both ATM UNI DTE and DCE Lports. The operation of the two Lport types is essentially the same, and the only differences between DTE and DCE are in the area of integrated local management interface (ILMI) processing. The user must use the ATM UNI DTE Lport type when configuring a feeder port for OPTimum cell trunk, OPTimum frame trunk, and FR NNI Lports.

When the user configures an ATM UNI DTE or ATM UNI DCE Lport, it must specify the number of bits that will be used for the ATM VPI and the ATM VCI fields when the address in the ATM cell is extracted. The values set for the VPI and VCI field lengths will determine the allowable VPI and VCI ranges for all ATM circuits configured to terminate on that DCE/DTE Lport (as well as for all OPTimum cell trunk, OPTimum frame trunk, and FR NNI Lports configured).

ATM UNI DTE and ATM UNI DCE Lports can terminate the following circuit types:

■ ATM circuits that originate from another ATM UNI DTE or ATM UNI DCE Lport in the network

■ FR-ATM service interworking circuits that originate on an FR UNI DTE or FR UNI DCE Lport in the network

6.7.2   Direct Cell Trunks

Direct cell trunks are an Ascend proprietary method for carrying circuit traffic between switches in a network of Ascend switches. Direct cell trunks are configured as point-to-point links between a pair of physically connected switches. Ascend's Virtual Network Navigator routing software is used to determine the topology of the switches in the network, and this information is used to route requested circuits through the network.

B-STDX software Release 4.2 adds the ability for the user to configure a direct cell trunk between the B-STDX and the CBX-500 ATM switch. Prior to Release 4.2, it was not possible to configure the B-STDX and CBX 500 to be part of the same routing domain, since there was no way for the two switch types to directly connect to one another and exchange routing information. This direct cell trunk connectivity allows the user to establish one contiguous network spanning both B-STDX and CBX 500 switch types.

Note: No other Lport may be configured on a Pport that has a direct cell trunk Lport configured. The direct cell trunk Lport does not require the prior establishment of an ATM UNI feeder port.

Direct cell trunks reserve VPI 0 for all circuits carried on the trunk. There are two control circuits established: one on VPI 0/VCI 16 for card-to-card communications, and one on VPI 0/VCI 17 for control-processor-to-control-processor communications. These control circuits are used for the transmission of routing and administrative information between the attached switches.

Direct cell trunks are integral to the Ascend concept of multiservice circuits. A multiservice circuit is any circuit that travels across multiple transmission technologies to get from the ingress point of the network to the egress point of the network. For example, in a mixed network of B-STDX and CBX-500 switches interconnected via direct cell trunks, a circuit can connect any two user interfaces regardless of whether the interfaces connected are frame relay or ATM. Also, the circuit can traverse both frame and cell trunks to connect the circuit endpoints. This multiservice circuit concept allows the user to ignore the underlying technology and simply configure the desired connectivity.

6.7.3   OPTimum Frame Trunks

The B-STDX supports the concept of using a VCC through an ATM core network as a logical trunk to interconnect switches. Essentially, this

OPTimum frame trunk is a proprietary version of FR-ATM network interworking, with the advantage that this Ascend proprietary trunk connection allows the switches to exchange management and control traffic. (The standard version of FR-ATM network interworking provides only an NNI interface, and no control and administrative traffic can be sent across this standard interface.) Since the OPTimum frame trunk rides across an ATM VCC, the user can configure multiple OPTimum frame trunks on a single physical interface. Each OPTimum frame trunk is an independent entity to the B-STDX, and the routing protocol is exchanged over each logical trunk separately. Figure 6.17 shows this concept graphically. In this example network, there are three separate OPTimum frame trunks configured across a single physical connection into the core ATM network.

When the B-STDX switch multiplexes multiple circuits across the OPTimum frame trunk, the switch will prepend an Ascend proprietary trunk header to the frame before segmentation and transmission. This header contains a circuit ID that provides the information the receiving switch needs in order to determine where to deliver frames that arrive over the ATM network.

As the name implies, OPTimum frame trunks can carry only frame-based data (even though the data are transmitted via ATM cells). Since OPTimum frame trunks tunnel through an ATM VCC, the B-STDX

Figure 6.17
ATM OPTimum frame trunk.

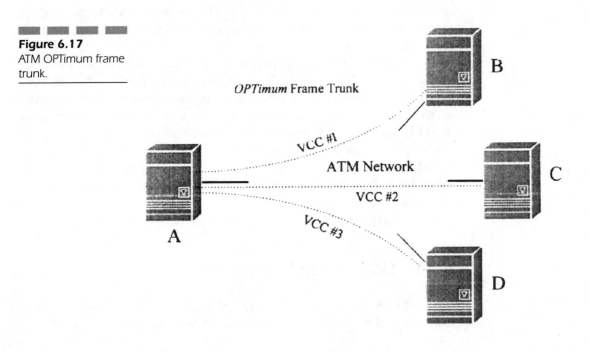

must apply the proprietary header for the destination switch to successfully demultiplex the traffic arriving over the trunk. For frame traffic, this header is applied as part of the AAL5 PDU. However, this header cannot be applied to native ATM cell traffic. Consequently, OPTimum frame trunks are restricted to carrying FR-ATM network interworking traffic and can originate/terminate only on B-STDX switches. It is not possible to terminate an OPTimum frame trunk on a CBX-500 switch or use this trunk type to forward native ATM cell traffic.

Note that the user must configure a "feeder port" before OPTimum frame trunk Lports can be configured. Since the OPTimum frame trunk Lports use VCCs through the ATM core, the feeder port must be an ATM UNI DTE Lport on the B-STDX. The user must reserve some small piece of port bandwidth for this ATM UNI DTE Lport, and the remainder of the port bandwidth will be available for use by OPTimum frame trunks.

6.7.4   OPTimum Cell Trunks

The B-STDX also supports a second type of OPTimum trunk, known as an OPTimum cell trunk. An OPTimum cell trunk is similar to an OPTimum frame trunk in that it allows the B-STDX to carve out a logical piece of bandwidth from a core ATM network and use this bandwidth as a logical trunk to interconnect switches. However, the major difference is that the OPTimum frame trunk tunnels across an ATM VCC, whereas the OPTimum cell trunk tunnels across an ATM VPC. Inside the VPC, the B-STDX assigns a unique VCC to identify each circuit traveling over the trunk. Figure 6.18 shows this concept graphically. In this example network, there are three separate OPTimum cell trunks configured across a single physical connection into the core ATM network.

As described above, OPTimum frame trunks require the switch to apply a proprietary header to all traffic transmitted across the trunk. OPTimum cell trunks do not need to apply this header, since they provide the ability to assign a unique VCI to each circuit that is routed across the trunk. The destination switch can then use the VCI to successfully demultiplex the arriving traffic. VPI/VCI identifiers are assigned beginning with VPI 1 VCI 32. [*Note:* VPI/VCI assignment will depend on how the user has chosen to segment the available address bits between the VPI and VCI fields. The VPI must be from 1 to n, where n is determined as $(2 \times \text{number of bits of VPI}) - 1$, and VCI starts at 32 up to $(2 \times \text{number of bits of VCI}) - 1$.]

Figure 6.18
ATM OPTimum cell
trunk.

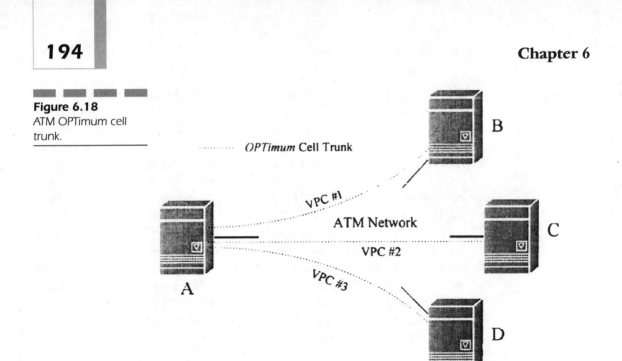

Figure 6.18
ATM OPTimum cell trunk.

This ability allows OPTimum cell trunks to support either frame-based interworking traffic or native cell-based traffic, and allows OPTimum cell trunks to originate/terminate on either B-STDX or CBX-500 switches. Along with direct cell trunks, OPTimum cell trunks can be used to connect the B-STDX and CBX 500 switches into one integrated routing domain.

Note that the user must configure a "feeder port" before OPTimum cell trunk Lports can be configured. Since the OPTimum cell trunk Lports use VPCs through the ATM core, the feeder port must be an ATM UNI DTE Lport on the B-STDX. The user must reserve some small piece of port bandwidth for this ATM UNI DTE Lport, and the remainder of the port bandwidth will be available for use by OPTimum trunks.

6.7.5 Network Interworking (FRF.5)

FR-ATM network interworking is defined as connecting two or more FR switches through an ATM core network. Based on the descriptions above, it is clear that OPTimum frame trunks and OPTimum cell trunks provide the ability to connect multiple B-STDX switches through an ATM core. Put another way, these OPTimum trunks are a proprietary form of FR-ATM network interworking that works only if the switches on both sides of the ATM cloud are Ascend equipment. However, the Ascend B-STDX also supports a standards-based FR-ATM network interworking solution that can be used to connect to switches from any manufacturer. According to the FRE5 specification for FR-ATM network interworking, this interworking approach uses a VCC through the ATM core network and treats this logical bandwidth as if it were an FR NNI connection between the two switches. Obviously, this approach requires that the switches be in separate routing domains, since there is no way to share administrative or routing information across an NNI boundary.

There are two flavors of standard FR-ATM network interworking: one-to-one interworking and many-to-one interworking. In one-to-one interworking, each FR circuit that crosses the FR-ATM boundary will require its own VCC through the ATM core. This will result in a larger number of ATM VCCs required and additional overhead, since the local management interface (LMI) must be run separately across each VCC. In the many-to-one interworking scenario, the switch is able to multiplex multiple FR circuits across a single ATM VCC.

For standard many-to-one FR-ATM network interworking, the B-STDX user can configure a single FR NNI Lport on top of an ATM UNI DTE feeder port. Note that the user will need to specify the VCI of the ATM VCC when this Lport is configured. Multiple FR circuits can terminate on this FR NNI Lport, and the data from these circuits will be multiplexed across the specified VCC. For one-to-one network interworking, the B-STDX user must configure one FR NNI Lport for each FR circuit that must cross the ATM network. The user then configures a single PVC to terminate on this FR NNI port.

6.8   Call Admission Control and Bandwidth Mapping

When a frame relay circuit is provided in an Ascend network that includes a B-STDX and/or CBX 500 core and one of the Ascend trunk ports, a call admission control (CAC) and bandwidth mapping function is performed.

Via the NMS, the operator specifies the circuit endpoints on the network by pointing and clicking on the relevant node, card, Pport, and Lport via the graphical user interface. The frame relay traffic contract is then provided in terms of CIR, Be, and Bc. Under the Virtual Network Navigator virtual circuit routing architecture, the terminating node (Ascend switch) with the highest node ID is designated as the source and is responsible for initiating the best-path calculation and call setup procedure. The values entered for the frame relay traffic contract are then translated to the ATM equivalent using a set of equations as follows:

$$PCR_{0+1} = (CIR + EIR) \times IOH/8$$

$$SCR_0 = CIR \times IOH/8$$

$$MBS_0 = Bc \times IOH/8$$

where IOH is the interworking overhead (N_Avg=256, so IOH=[(N_Avg+AALHS)/48]/N_Avg=0.0219.) The values of PCR and SCR are then passed to the CBX 500 call admission control scheme, which determines if there is sufficient bandwidth to meet the QoS requirements of the circuit. (A complete description of the Ascend CAC scheme is given in a separate document available from Ascend Product Marketing.)

For service interworking PVCs that terminate on a B-STDX ATM UNI DTE interface, the B-STDX treats these circuits as non-real-time variable bit rate (nrt-VBR) PVCs on the ATM interface. The ATM traffic descriptors for FR-ATM service interworking PVCs are calculated and are displayed to the operator as a guide for providing the separate ATM network. However, when service or network interworking is implemented via a multiservice PVC over Ascend trunks (any combination of direct or OPTimum cell trunks) in an all-Ascend network, the B-STDX will offer the user more choices regarding how to treat the circuit in the ATM portion of the integrated network. Specifically, the following options are offered:

Available bit rate (ABR).   This implies that the circuit will be allocated BW=MCR=CIR and will have access to the UBR/ABR buffer on switches that do not have the CBX 500 FCP card installed. (When the FCP is present, the interworking circuits benefit from the additional buffering provided.) The MCR for this circuit is calculated from the CIR of the FR network according to the following equation:

$$MCR(0+1) = \frac{CIR \times OH_B(n)}{8}$$

This equation can be derived in the same manner as the method 1 equation for SCR above, except that it is based only on the CIR, not on the sum CIR + EIR. This option provides the most accurate model for the FR data while passing through the ATM portion of the network, since it provides a guaranteed bandwidth (derived from the guaranteed FR CIR) while allowing the user to burst higher based on available network bandwidth. This option is the default option when configuring a multiservice PVC.

Unspecified bit rate (UBR). This implies that the circuit will be allocated zero bandwidth and will receive best-effort service. This option is useful for frame relay PVCs defined under a zero CIR service.

Variable bit rate—non-real-time. This implies that the circuit will be allocated some bandwidth according to the chosen CAC algorithm, which makes use of the converted traffic descriptors. This choice offers the flexibility of getting the CLR guarantees if they are of interest.

When the user configures a multiservice PVC, the desired traffic engineering option can be configured and the network will automatically account for the effective bandwidth and treat the PVC appropriately.

6.9 Traffic Shaping

Traffic shaping is provided to give some measure of control over frame relay traffic bursts.

The ATM traffic parameters are used to define a traffic contract that defines the particular bandwidth and burst requirements of a given virtual circuit. These parameters include PCR, SCR, and MBS. Traffic shaping, which occurs on egress at the point of transmission, ensures that the cells emitted are in conformance with the ATM traffic parameters being enforced by the receiver's UPC function. Traffic shaping is a method of flow control that will limit the amount of traffic transmitted so that it will conform to the receiver's parameter settings, thereby reducing the likelihood of lost data along the transmission path.

The ATM CS and ATM IWU cards perform a traffic shaping function using the NEC µPD98401 ATM segmentation and reassembly (SAR) chip. Cell bypass traffic has no traffic shaping applied, as it does not pass through the ATM SAR.

The ATM CS and ATM IWU cards provide support for 16 discrete traffic shapers. Each shaper is independently configurable for a given combination of PCR, SCR, and MBS. The B-STDX software reserves the first shaper as a "line rate" shaper, with a profile of PCR=SCR=line rate. The remaining 15 shapers are configurable by the user at the Pport level. A given shaper is assigned values for PCR, SCR, and MBS and a priority from 0 to 15.

At provision time, the user enters values for CIR, Be, and Bc to define the frame relay traffic contract. The NMS calculates the equivalent ATM traffic parameters (PCR, SCR, and MBS) using the following set of equations, which are based on the ATM Forum B-ICI Appendix A, Method 1 definitions, and displays them to the operator.

$$PCR_{0+1} = (CIR + EIR) \times IOH/8$$

$$SCR_0 = CIR \times IOH/8$$

$$MBS_0 = Bc \times IOH/8$$

where IOH is the interworking overhead (N_Avg=256, so IOH=[(N_Avg+AALHS)/48]/N_Avg=0.0219.

- *Service interworking circuits that terminate on a B-STDX ATM UNI.*
  The user is presented with a pick list that displays the configured traffic shapers. Note that the PVC values for PCR, SCR, and MBS are not used in this process and are presented to give the user an idea of what shaper should be selected by the user in the configuration process. Once the operator has selected a shaper, all data associated with that virtual circuit are mapped to that shaper. As data are transmitted, the shaper will meter out the traffic for the virtual circuit so that it will conform to the PCR/SCR/MBS profile of the relevant shaper.

- *Network interworking circuits over OPTimum frame trunks.* The entire frame trunk is shaped according to the shaper selected for the trunk when it is created.

- *Network interworking circuits over OPTimum cell trunks or direct cell trunks.* Effective with the OPAL maintenance release, the shaper bank may be logically partitioned into three groups of five shapers per group. Each shaper bank will be associated with a QoS class (rt-VBR, nrt-VBR, and UBR). When a circuit is to be routed over a cell trunk, the circuit setup procedure will use the QoS class of the circuit, which is carried in the circuit setup message, to select a shaper group. Once the shaper group is selected, the circuit manager will match the SCR (calculated from the CIR if frame relay traffic parameters are used) with a shaper SCR within the

QoS group and will select this shaper as the one to use on the cell trunk output. The match will be done in such a way that the selected shaper will always have a configured SCR that is greater than or equal to the SCR of the circuit. This mechanism will be available only for the ATM CS and ATM IWU cards. The NEC chip on the card provides the shaping function. Each shaper has a priority assigned to it during configuration. The NEC chip schedules shapers on a priority basis, so that a higher-priority shaper will service its circuits before a lower-priority shaper. Using this capability, you can prioritize the shaper groups so that circuits assigned to rt-VBR can have a higher priority (for transmission) than circuits in the nrt-VBR and UBR classes. If no prioritization is desired the shaper priorities can all be set to the same value, and then all shapers will be serviced equally.

■ *Service interworking circuits over direct cell trunks.* These are shaped in the same way as for network interworking circuit.

6.10 Addressing

The ATM CS and ATM IWU modules provide support for 12 bits of address space. The user may configure the card with respect to how bits are assigned to virtual path and virtual channel identifiers.

When connecting the B-STDX to a CBX 500 ATM switch using these modules, the trunk protocol passes the number of bits of VPI/VCI currently in use on the two cards that terminate the trunk. The cards choose the minimum number of bits for VPI and VCI on the endpoints so that each endpoint will agree on the number of bits for both VPI and VCI, and will know the limit to the number of circuits available.

6.11 ATM Quality of Service

The ATM CS and ATM IWU provide support for ATM QoS classes. A multiqueue output buffering scheme is employed to provide dedicated resources to each QoS class (see Fig. 6.19).

Queues are scheduled using a priority servicing scheme that guarantees optimal performance for CBR ATM virtual circuits. When passing frame relay traffic to ATM across an Ascend trunk facility (direct cell

Figure 6.19
Output buffering scheme for ATM CS and ATM IWU modules.

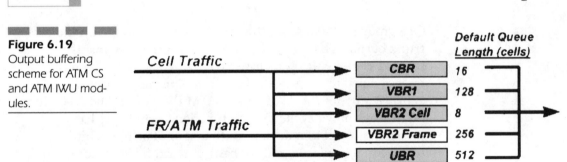

trunk, OPTimum cell trunk, or OPTimum frame trunk) to a CBX 500 ATM switch, the operator may choose to map to any of the non-real-time traffic classes supported by the CBX 500: nrt-VBR, ABR, or UBR.

6.12 Summary

Frame relay–ATM interworking technology plays an essential role in building scalable broadband network architectures. Effective interworking implementations address two key requirements for public data networks:

- *Scalability.* ATM equipment switches are available today that support trunking speeds as high as OC12/STM4 (622 Mbps). Interworking technology allows frame relay networks to extend their scale of deployment by leveraging these high-capacity infrastructures.

- *Migration path.* Frame relay–ATM service interworking technology provides a migration path for data service customers from frame relay to native ATM services.

Ascend Communications, Inc., has extensive experience with the real-world application of frame relay–ATM interworking technology in some of the world's largest public data networks. A family of ATM modules for the B-STDX multiservice platform provides a flexible implementation for a number of networking applications. When applied in combination with sophisticated circuit routing protocols, operations costs can be dramatically reduced as a result of automated provisions that make the technical intricacies of the interworking function transparent to the network operator.

Enterprise Networks and the Internet: Carrying IP over ATM

7.1 Need for IP Support

One of the key considerations for ATM technology in recent years has been the support of IP. This requirement is driven by (1) the desire to support the embedded base of applications and enterprise networks (including intranets) and (2) the desire to have access to the Internet, including virtual private networks (VPNs) over it. Beyond basic support of IP over ATM, the industry has looked at ways to use the advantages of ATM to simplify IP-level Layer 3 protocol data unit (PDU) forwarding. In view of the increased corporate dependence on information, including data, video, graphics, and distributed resources (Web access), users and planners want faster, larger, and better-performing networks—namely, higher speeds, scalability, and better performance and management. Specifically, companies need all-points broadband networks, interconnecting major corporate locations and remote branch locations. ATM has the potential for meeting these customer expectations when used in combination with IP and routers. Others are advocating entirely new approaches, e.g., the Resource Reservation Protocol (RSVP).

This chapter examines a variety of issues related to enterprise network migration and the use/support of ATM to bring about design improvements. Some of the ideas presented may be controversial. Some of the available approaches could entirely eliminate the use of routers (e.g., end-to-end ATM with IP route selection in the end system); other approaches use ATM as a "fat pipe"; other approaches either diminish the dependence on hop-by-hop routing or make better use of routing by introducing new paradigms; other approaches are more conventional and make little use of ATM. These last are likely to be short-sighted and short-lived as we enter the new millennium.

There is an obvious use of ATM to connect dispersed routers via a mesh of permanent virtual connections (PVCs) using multiprotocol encapsulation. This entails upgrading the network-side hardware of the routers with an ATM access board that supports encapsulation of IP PDUs, segmentation, and basic ATM features, such as cell generation, cell relay, and possibly traffic shaping, signaling, and maintenance cell generation. This is illustrated in Fig. 7.1. However, this still requires conventional routers in support of the enterprise network. Here the glass can be seen as half full or half empty. Conservative planners may breathe a sigh of relief that the existing enterprise architecture can be kept intact, and the router simply has a "bigger/better WAN pipe" (glass half full). More aggressive planners, on the other hand, may dislike the burden, in

Figure 7.1a
Enterprise networks composed of dedicated lines, and the difficulty of adding a new site to the network.

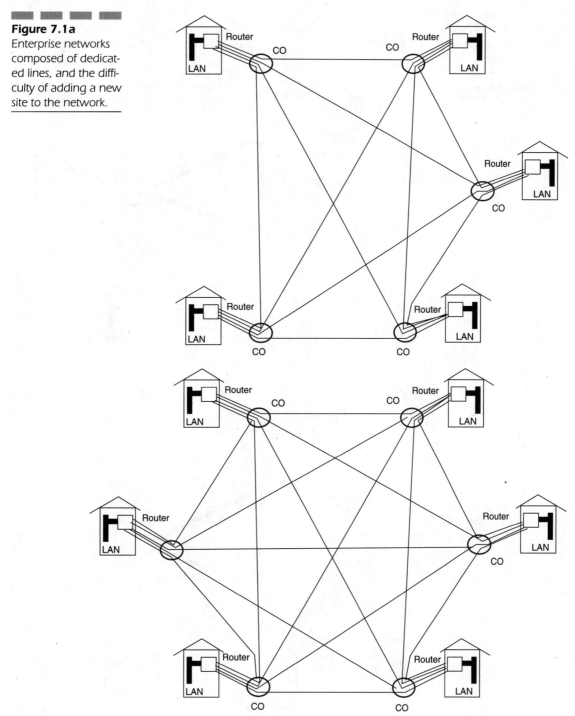

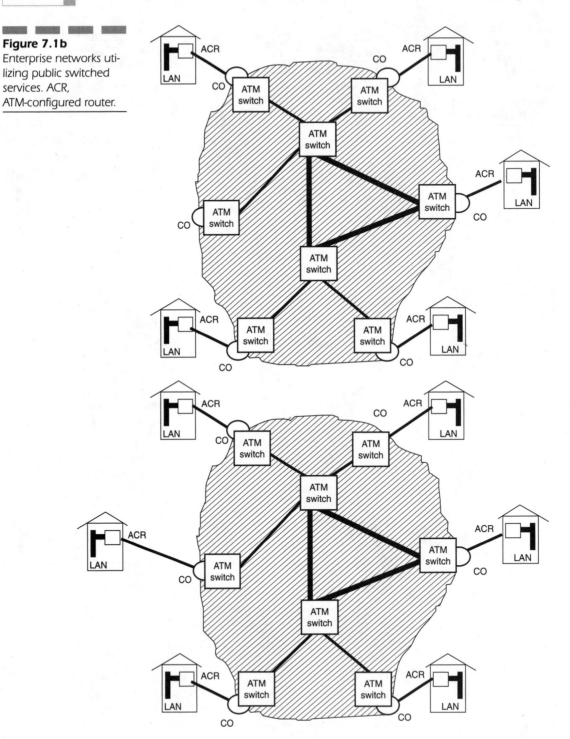

Figure 7.1b
Enterprise networks utilizing public switched services. ACR, ATM-configured router.

this usage of ATM, of having to retain traditional and expensive routers (the glass is half empty). Hence, alternatives are being sought.

Radical redesigns that reduce the number of IP subnets and nearly eliminate the obligatory use of routers, and thereby have the potential for greatly reducing costs, are possible. However, this approach is not given much priority by people with vested interests in routing technology. Therefore, work is underway to address the issue of deploying all-points broadband networks in a corporation while retaining routers.

Work on ATM use in IP networks is being undertaken by the ATM Forum and by the Internet Engineering Task Force (IETF), sometimes in cooperation, sometimes in competition. Issues under discussion include the following:

- Cost-effective deployment of broadband connectivity to all corporate locations
- Quality of Service (QoS) [Resource Reservation Protocol (RSVP) by IETF; UNI 4.0 signaling by the ATM Forum]
- Traffic management (Classifier/Scheduler/FlowSpecs by the IETF; Traffic Management 4.0 specification and traffic contracts by the ATM Forum)
- Addressing [Flow IDs and protocol-independent multicast (PIM) by the IETF; VPI/VCIs and Next Hop Resolution Protocol (NHRP) by the ATM Forum]
- Routing (IGP/EGP by the IETF; PNNI by the ATM Forum)
- Interneting [Routing Over Large Clouds (ROLC) and NHS by the IETF; multiprotocol over ATM (MPOA) and I-PNNI by the ATM Forum]

Some of the challenges of using routers in an ATM environment that are being addressed are [mc]

- Support for more than best-effort and constant bit rate services on the same router-based network
- Support for best-effort service over ATM (which service class to use, and whether these classes are supported by a specific carrier or carriers)
- Support for circuitlike service via IP on a router-based network
- Support for service levels regardless of physical media and network discipline (e.g., ATM/LAN switching, traditional IP routing, IP routing over ATM)

- Including a way to request QoS, a way to deliver it over various lower layers, and a way to find routes that support the needed QoS on a router-based network

- Making the new designs/technologies easier to deploy in router-based networks

When Ipsilson Networks introduced the IP Switching concept in 1996, the announcement attracted a lot of attention, although the ATM Forum and the IETF were already working on ways to route IP over ATM. Ipsilon offered the argument that emerging standards were too complex, that interoperability and scalability might suffer, and that its IP Switching was simpler and more robust. A number of vendors now support IP Switching, including Digital Equipment Corp., General DataComm, Hitachi, Ltd., ECI/Telematics, and NEC Corp. Cisco Systems, IBM, Cascade/Ascend, Toshiba, and 3Com have come up with alternative solutions based on similar principles. By press time, the idea of generic IP switching, the forwarding/routing function that seeks to directly and cohesively switch and forward frames without having to do Layer 3 processing unless absolutely needed, had become well accepted. Other vendors, such as Cisco, promote their own "optimized solution" (e.g., Tag Switching/NetFlows for large internetworks), but still support MPOA for smaller networks. The industry is now working toward a standard [Multiprotocol Label Switching (MPLS)]. These issues are expanded upon in the sequel.

7.1.1 Key Motivations

There is a clear recognition that the router-based networks of the late 1980s and early 1990s do not scale well as the number of points increases. The fundamental problem at hand is simple to state, but the jargon is overwhelming and the solutions offered at present are often parochial. The problem is, how does a network planner interconnect (say) 1000 users who (1) need a considerable amount of bandwidth because of evolving graphics/video-based applications (QoS support), and (2) may be geographically dispersed?

We started out in the mid-1980s with a need to create small Layer 2 groups (what have been called bridge groups) because of the performance restrictions of shared media LANs, including the generation of broadcast traffic across related communities of users. Because of the performance restrictions, network planners may have formed, say, 20 groups of 50 users each. The feudalistic fragmentation at Layer 2 was reflected in a similar fragmentation at Layer 3. Network planners soon discovered plausible arguments to support this fragmentation. Technologists soon provided explanations along the following lines: "Isn't it obvious that the Marketing Department should have a separate subnetwork from the Support Department?" "Isn't it obvious that the centralized East Coast Support Department should have a separate subnetwork from the colocated West Coast Support Department?" "Isn't it obvious that the Academic Department should have a separate subnetwork from the Administration Department?" "Isn't it obvious that the Small Cars Manufacturing Department should have a separate subnetwork from the Medium-size Cars Manufacturing Department?" *The answer is absolutely no!* Why are separate subnetworks needed at the business level? There is no reason that the East Coast Support Department should keep separate data from the West Coast Support Department (unless they are geographically separate), because at some point, all of this information has to be merged, reported, and acted upon. Subnetworks are, however, also needed to limit broadcast domains of the underlying broadcast media. But new switched Layer 2 media now allow control of these broadcast domains.

Once this myriad of subnetworks has been created, there is an a posteriori need to provide a connection among the various subnetworks. Enter the need for routers.

What is required now of corporations that want to save money in their networking budgets is to recognize that the segmentation was a necessity imposed by the immaturity of the 1980s technology, not a sacrosanct overall architectural imperative or an untouchably elegant construct of perennial merit.

The deficiencies of the technology of the 1980s that originally gave rise to the need for routers have been overcome, and corporations that want to take advantage of these technological improvements can reduce their costs as much as three-fourths. Instead of spending $80,000 on a router for each core node, they can get the same functionality from a $20,000 switch. Routing can then be relegated to the edges.

In the early 1990s, venerable mainframe manufacturers were not telling their customers, or the world, that a much cheaper client/server (or now WWW) technology could replace their mainframes. Similarly, in the late 1990s, venerable router manufacturers are not telling their customers that a much cheaper technology could replace their routers, at least in the core of the network. Naturally, for a corporation to gain the computing benefits of client/server, it had to change the architecture, redesign its applications, and deploy new hardware, but the savings were significant and well worth it. It should be no surprise that to gain the networking benefits of pure switching-based ubiquitous broadband communication, a corporation has to change the architecture, redesign the network, and deploy new hardware, but the savings are significant. So, the router vendors are right: You cannot just throw away the router; you have to first redesign the network to eliminate the need for routers at every core node, then secure the benefits[1] [dmsw].

In contemplating the significant network hardware, network support, and network operations savings accomplishable with switching, corporations should appreciate the concomitant need for new architectural creativity. Corporations that embrace switching in the core and routing edge-to-edge will be the ones with reduced networking burdens in the coming decade. It should also be noted that previous efforts to secure connection-oriented services over a connectionless technology have failed—for example, IVDLAN (integrated voice/data LAN, IEEE 802.9) and FDDI II. Who's to say that new efforts aimed at establishing QoS-based all-points broadband connectivity, which is a clear corporate requirement, without using a connection-oriented technology will surely succeed in the market?

7.1.2 Why Routers Were Needed to Begin With in Enterprise Networks

In the mid-1980s, designers of (then small) data networks needed a service that enabled them to

1. Forward IP PDUs toward the destination (in an effective manner)

2. Provide reliable concatenated-link-by-link connectivity

---

[1]The IP "routing function" (i.e., path selection) can be undertaken in software in the end system (e.g., PC), and the desired path can be identified completely and uniquely at the edge of the network. The topology map to support network reroutes would be retained at the Data Link Layer rather than at the IP Layer; this implies a shift in responsibility from the user/organization to the network itself. The whole concept of addressing needs to be revisited. In the beginning telecommunication networks did not support the ability to address remote locations in real time (except with circuit switching, but this is too slow and does not support enough bandwidth). Hence, IP introduced a way to accomplish that—being able to identify the location within that realm was part of the required machinery. But now ATM-based networks can support real-time addressability of devices attached to the network. Hence there is a redundant function of Layer 2 and Layer 3 addressing. Why should redundancies be retained, in fact institutionalized? There are opportunities for savings if the redundancies are eliminated.

3. Have a knowledge of the topology of the network, in order to accomplish point 1

4. Support point 1 by introducing an addressing mechanism at Layer 3 that enabled the sender to pipeline data in real time (without appreciable setup latency) to any entity on the network (i.e., to support any-to-any communication, rather than just fixed point A–to–fixed point B communication)

No such service was available in the mid-1980s; therefore, these designers developed a box that performs all these functions. They invented routers. Routers forward IP PDU, collect and maintain topology information, and support reliable concatenated link-by-link connectivity, but they do so by recomputing the IP decision at each endpoint of a Layer 2 link. This can be inefficient for a number of reasons:

1. Each router must have a complex apparatus to generate, aggregate, distribute, and verify topology information.

2. Each router must process PDUs through the physical, data link, and IP protocol engines. This becomes a bottleneck.

Now the challenge is to relegate routers to just job function 1 described above. In particular, there has to be a decoupling of IP processing and the movement of information along a trajectory close to the destination. Some analogies might help the reader understand the issue of "routing once switching many" anecdotally.

1. An overnight service wants to move a package from customer location A (to an aggregation point X in a city, to a major distribution node Y in a central location in the United States, to a fan-out point Z in a remote city) to recipient location B. The service provider naturally wants to determine if a truck route, an airplane route, or some other route is appropriate (perhaps A is in New York City and B is in Newark, N.J., and so a truck route may be better). Also, the service provider wants to be aware of the topology (e.g., airline H is on strike, so use airline K). However, the service provider would not want to have to *open* the package at points X, Y, and Z in order to be in a position to move the package toward its destination. Yet, when moving data, routers at each end on a link must process the IP PDU (the analogy of opening the package).

2. A driver enters the New Jersey Turnpike at Entrance 1, and crosses into Delaware on his or her way to Washington on Highway 95. The driver knows that Baltimore is 60 miles away. In Baltimore, the driver is planning to get off and take the Baltimore-Washington Expressway. Now, just after leaving Wilmington, Delaware, the driver does not want to feel obligated at every exit coming up (perhaps one every couple of miles), to read in detail the "branch off" instructions on the large green signs, and make a stay/get-off decision (the analogy of having to route an IP PDU at a router site along the path). The exit of interest is 60 miles away, and so the forward projection of the car should not be tied to a complex decision at every upcoming exit; rather, the decision should be "just go straight for about 50 minutes, then start looking for signs."

3. A mail carrier is required to deliver mail to 1000 business buildings on his or her itinerary. The mail carrier would prefer to drop off the mail on the ground floor (Physical Layer), say in 30 s. Instead, the mail carrier must queue up and take the elevator to the second floor (Data Link Layer), get off, and then take another elevator to the third floor (Network Layer). Say this takes 2 min. After the mail carrier hands off the mail to a clerk on the third floor (say 30 s), he or she must wait for the elevator down to the second floor and queue up for another elevator to the first floor. This could also take 2 min. Similarly, it would be better if the data did not have to be taken up to the Network Layer in order for a forwarding decision to be made.

These days, unlike the mid-1980s, there are communication services that can support functions 2, 3, and 4 above (in some cases even 1). In effect, one wants to push the routers to the edges of the network in order to provide an interworking function with a legacy application, but not have them in the core of the network to handle basic heavy-duty (industrial-grade) data forwarding functions, since they are no longer required there (so that one can use a $20,000 switch instead of an $80,000 router inside the network). Topology maintenance and calculation can also be centralized (as one has seen with the route servers in the MPOA model) to eliminate having to replicate this intelligence at each hop-end—questions of reliability can be addressed just as they have been for decades with the Common Channel Signaling System No. 7 and in the 800 translation databases. Figure 7.2 depicts a first stage of the desired migration. Here routers have been relegated to the edges of the network. Some even claim that IP processing could, in theory, be relegated to the endsystem (software routers) and the network infrastructure be all ATM (this, however, would relegate ATM to the desktop).

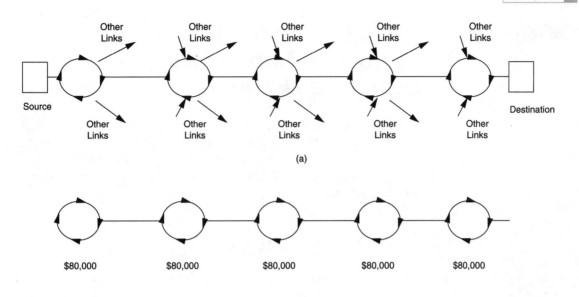

(a)

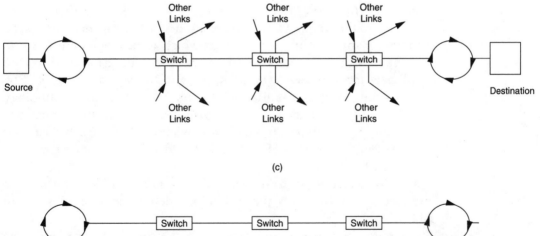

$80,000 $80,000 $80,000 $80,000 $80,000

(b)

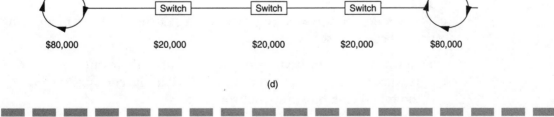

(c)

$80,000 $20,000 $20,000 $20,000 $80,000

(d)

Figure 7.2
Advantages of using ATM. (*a*) Physical layout today. (*b*) Cost today. (*c*) IP PDU forwarding done only at the edge of the network; routing (topology discovery/maintenance) functions done either as usual between routers, or via centralized route server. (*d*) Reduced cost.

7.2 A Baseline for Discussion: Dynamics in Enterprise Networking Designs

New major market forces are at work in the enterprise backbone as well as at the desktop/hub level. At the desktop level, new solutions seem to be emerging all the time; however, where the real expenditures are, namely at the backbone and WAN level, new solutions are hard to come by and/or implement.

7.2.1 Internetworking Level

How can planners make some sense out of all the available proposed alternatives to streamline routing? Are these alternatives real and fundamental, or are they just vendor-advanced variations of the same radical or reactionary solution? Observers note that "there is widespread excitement on Wall Street, while at the same time there is confusion on Main Street." Clearly, there are new trends in networking. So corporate planners must try to guess which technologies are likely to be winners and which losers. R&D users in a corporation tend to look at (all) new technologies. Eventually, these technologies are adopted by technical/power users. Eventually all users will dabble in a technology, on the assumption that it makes it to success [some technologies, such as dual queue distributed bus (DQDB), switched multimegabit data service (SMDS), packet-over-ISDN, FDDI II, and IEEE 802.9, never made it to market]. See Fig. 7.3.

It is always important to understand the embedded base. At this time one predominantly finds the technology deployment highlighted in Table 7.1.

The following are well-recognized trends of the late 1990s, modulating the embedded base.

■ Apparent desire to retain widespread use of fairly mature technologies such as Ethernet and TCP/IP in the end-system space (hosts and desk-proximity networks).

■ Desire to upgrade rather than replace (e.g., upgrade 10-Mbps Ethernet to switched Ethernet; upgrade 10-Mbps Ethernet to 100-Mbps Ethernet).

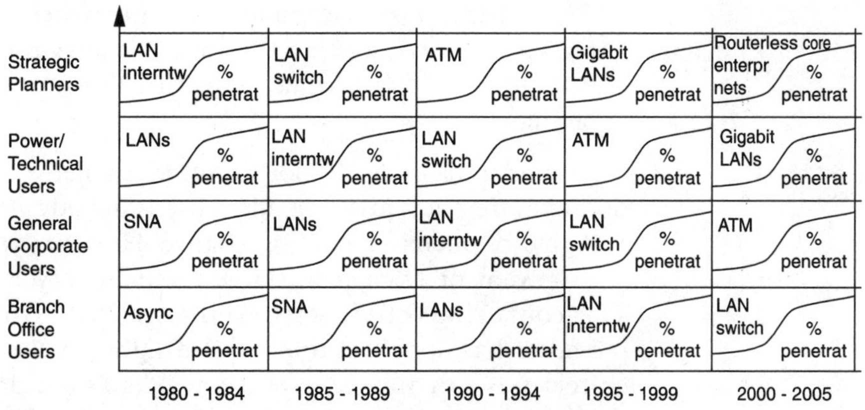

	1980 - 1984	1985 - 1989	1990 - 1994	1995 - 1999	2000 - 2005
Strategic Planners	LAN interntw % penetrat	LAN switch % penetrat	ATM % penetrat	Gigabit LANs % penetrat	Routerless core enterpr nets % penetrat
Power/ Technical Users	LANs % penetrat	LAN interntw % penetrat	LAN switch % penetrat	ATM % penetrat	Gigabit LANs % penetrat
General Corporate Users	SNA % penetrat	LANs % penetrat	LAN interntw % penetrat	LAN switch % penetrat	ATM % penetrat
Branch Office Users	Async % penetrat	SNA % penetrat	LANs % penetrat	LAN interntw % penetrat	LAN switch % penetrat

Figure 7.3
Promulgation of new technologies in the networking chain.

- Low price (e.g., less than $100 per desk).
- Businesses' critical dependence on computing and communication.
- Consolidation of equipment vendors, carriers, and Internet service providers (ISPs).
- Operations and network management expenditure concerns.

TABLE 7.1

Embedded Enterprise Network Technology, Late 1990s

PCs	Intel (Pentium II)
Desktop OSs	Microsoft Workgroup/Windows 97/NT
Enterprise connectivity	Physical: NICs/switched Ethernet/100-Mbps Ethernet (e.g., 3Com)
	Logical: TCP/IP
	Applications: Web/intranet/Internet (e.g., Microsoft, Netscape)
Servers	Unix, NT
Internetworking	Router-based technologies (e.g., Cisco)
Access	SoHo networking: dial-up, frame relay (e.g., Ascend)
	Internet access (e.g., Ascend)
Carriers	LECs, CLECs, IXCs, ISPs

■ An avalanche of new technologies that may well confuse the
market. Vendors quote fast time to market as a must, since 50
percent of the profits come in the first 12 months of a new
product's life.

Corporate planners do not really design networks from the ground
up anymore; they just upgrade what is already there. Hence, there is
a degree of inertia regarding change. However, some radical change
(e.g., movement from mainframe to client/server) may have benefits.
The confusing, often self-serving message from the technology
providers has done nothing to shed light on the fact that newer architectures can indeed save money. Table 7.2 depicts some of the
evolutionary "half-steps" in the right direction. The most critical, rev-
olutionary change would be to redesign the network to operate without core routers. Just making a router operate faster does not alleviate
the inefficiency of the traditional way data are carried across an
enterprise network. In the overnight mail example above (Example 1),
one does not want to unwrap each package at each stage, no matter
how dexterous the clerk is. In the highway example above (Example 2),
one does not want to be forced to look at every sign along the highway, no matter how fast one can read the sign (e.g., the sign containing just a short string that can be rapidly parsed). In the mail carrier
example above, making the mail hand-off on the third floor take 20
rather than 30 s, is not an elegant solution. Having to read each PDU
and make an individual decision is a brute force method of solving
a problem; it is like doing an exhaustive search in an optimization problem, rather than having a rapidly converging analytical
algorithm.

7.2.2 WAN Level

Technologies have been more stable on the WAN side than on the LAN
side. Specifically, frame relay has been commercially successful in the
past decade, proving the utility of a switched service. ATM is viewed as
being a couple of years behind frame relay. Some of the factors that have
helped frame relay are as follows:

■ Early support by canonical networks of the late 1980s, namely,
Systems Network Architecture (SNA)

TABLE 7.2

Evolutionary Steps

Requirement	LAN Approach	WAN Approach
Higher speed	Use switched Ethernet, 100-Mbps Ethernet, Gigabit LANs	Use ATM (frame relay not enough)
Dedicated bandwidth/ improved performance	Use switched Ethernet	Use service with QoS guarantees (such as ATM)
Multimedia	Use Isochronous Ethernet or higher-speed LANs	Use ATM
Improved management	Use virtual LANs	Use TLS (transparent LAN service)
Multicast and reservation	Use IP multicast, RSVP	Use IP/RSVP over ATM

- Generally a software upgrade of routers
- Existing access mechanisms (e.g., T1 tails)
- Support of a range of speeds as needed at the time (e.g., 56 kbps)
- Economic savings, compared to meshes of dedicated lines

Now there is a need for speeds higher than T1, but some of the same principles are at work. Table 7.3 depicts some transitions in views as related to frame relay.

One of the reasons frame relay was so successful is that it supported a move toward all-points connectivity; also, it enabled data to be relayed between points in the network without necessarily requiring hop-by-hop routing. Where it fell short was in supporting broadband and in supporting QoS (although some are now proposing ATM-like service classes for it).[2] ATM is the broadband WAN technology of choice at this time, particularly because it supports QoS.

There are clear expectations that TCP/IP-based applications should run well over ATM as the Data Link Layer technology. The problem is that there is duplication of addressing and routing functions between these two layers, and, further, that the LAN emulation (LANE) and classical IP over ATM (CIOA) approaches hide ATM's QoS aspects. The latter is a concern as related to the support of multimedia. Also, there is increased interest in supporting multicast communication.

[2]Even if specifications were published by the Frame Relay Forum, it is not clear that manufacturers would develop hardware—perhaps frame relay over ATM is a pragmatic compromise.

TABLE 7.3

Views on Frame
Relay

Early 1990s View	Late 1990s View
Limited in speed, does not scale	In spite of speed limitations, it will continue to be important
No QoS support	Vendor extension will include QoS
Data only	Voice now also carried in standardized form
Frame relay a feeder to ATM	Frame relay will be carried over ATM
Full upgrade to cell-based technology	Eventual upgrade to cell-based technology

The plethora of technologies and choices has led (in the late 1990s) to a hybrid situation, in which enterprise networks either continue to use routers in a traditional manner, or use routers connected to ATM switches and/or public ATM networks, or use routers connected directly to SONET-based private lines, or use ATM switches that run router software (e.g., Ipsilon's IP Switching). Table 7.4 depicts a late 1990s view of the two technologies.

The path to higher speed is generally, although not exclusively, via cell switching, whether directly at the ATM level or via Network Layer switching technologies. The exception is for the local environment, where Ethernet switching at the 10-, 100-, or 1000-Mbps level is often based on frame technology.

The current commercial outlook (brought about by vendors with vested interests) is that ATM will not operate end-to-end (desktop-to-desktop), but will be focused in the WAN backbone for broadband applications. One can think that this use is consistent with the way the technology was originally conceived in the late 1980s (frame relay will also continue to play a useful role for the foreseeable future). Must there be a single end-to-end technology? Not necessarily, although this would clearly be the ideal. However, this misses the point—specifically, the point that ATM could reduce overall networking costs. The issue of the value of ATM end-to-end is more related to the ability for end-system routing (i.e., being able to automatically select an end-to-end path right at the source) and a single address space in the WAN than to the speed at the desktop. Some vendors have focused attention on the lack of need for broadband to the desktop or on the availability of other desktop solutions (such as switched Ethernet, 100-Mbps Ethernet, or Gigabit Ethernet).

Some claim that the Internet will be the new "superhighway." It is worth noting that internal to the Internet backbone, ATM either is already being used or will soon be used by the key backbone providers, at least as a "fat pipe" technology. VPNs at the IP level may, in the core, use ATM. xDSL technology is receiving a lot of attention, but the penetration is still extremely low. Nonetheless, as more xDSL services are deployed, giving users access links operating at 1.5 or 6 Mbps, the need for high-capacity backbones will be accentuated. The need for ATM will be even more accentuated.

TABLE 7.4 Late 1990s View of Switching and Routing Technology

Feature	Switches (Original)	Switches (More Recent)	Routers (Original)	Routers (More Recent)
Connections	PVCs	PVCs and SVCs	None	Flows
Unit of processing	Cells	Cells, FR frames, LAN frames	LAN frames	Cells, FR frames, LAN frames
Traffic control	Service/QoS guarantees for CBR, VBR	Service/QoS guarantees for CBR, nrt-VBR, rt-VBR	Best effort	QoS support
Cost	Low	Low	High	Improving
Implementation	Hardware	Hardware and software support of functions (e.g., MPOA, LANE)	Mostly software	More hardware-based support
Scalability	Good	Better (with MPOA)	Somewhat limited	Improving
Speed	Single-digit gigabits per second	Double-digit gigabits per second	Megabits per second	Near gigabits per second
Standards	Open, ATM	Open, IETF	Proprietary	More open (e.g., OSPF, MPOA/MPLS)
Customers	Telcos	Carriers, ISPs, large users	ISPs, corporate users	Carriers, ISPs, corporate users

7.2.3 LAN Level

At the LAN level, Ethernet switching and 100/1000 Mbps Ethernet will be the technologies of choice for the foreseeable future. LAN switches are replacing hubs as the first connectivity tier. Switches support multiport bridging. The majority of LAN switches will be Ethernet frame-based rather than cell-based. Higher throughput will be obtained by higher-speed LAN technologies (rather than new technologies). Virtual LANs (VLANs) in their current manifestation will play only a minor role, according to observers. The early 1990s view that Ethernet was insufficient for new applications (including multimedia) and that it did not scale has been replaced as a result of its switching and speed extensions. Fast Ethernet costs about the same as Ethernet at the network interface card (NIC) level, and only twice as much at the hub level. Switched Ethernet hubs now cost the same as regular hubs did in the mid-1990s. Switched Fast Ethernet is expected to be common by the late 1990s. The cost of Gigabit Ethernet will be about twice that of Fast Ethernet, but with a 10 times increase in performance.

Fast Ethernet went from research to standard in one year, and it provides 10 times the performance for twice the price (at $100 per NIC). Note that after only 1 year of technology maturity, Fast Ethernet costs half as much as ATM NICs with 5 years of technology maturity.

7.3 Methods of IP Support

The previous section described some of the issues, opportunities, and potential strategic directions for the deployment of all-points corporate broadband networks. This section describes the tactical approaches to IP support over ATM.

7.3.1 LANE

As discussed in Chap. 3, the ATM Forum's LANE specification defines how existing applications, e.g., IP-based LAN-situated applications, can operate unchanged over ATM networks. It also specifies how to communicate between an ATM internetwork and Ethernet, FDDI, and token ring LANs.[3] LANE is a logical service of the ATM internetwork. In this translational bridging interworking environment, Ethernet and token ring frames that carry IP PDUs (or for that matter, other PDUs) can cross an ATM network (in a segmented fashion) and be delivered transparently to a similar legacy network at the receiving end. Furthermore, a user on an ATM device can send information to an Ethernet or token ring device. LANE provides users with a migration path from pre-ATM architectures, without passing through successive stages of large-scale reinvestment.

LANE is an ATM-based internetworking technology that enables ATM-connected end stations to establish Medium Access Control (MAC)–Layer connections. It allows existing LAN protocols, such as Novell NetWare, DECnet, TCP/IP, MacTCP, or AppleTalk, to operate over ATM networks without requiring modifications to the application itself. LANE provides

- Data encapsulation and transmission
- Address resolution
- Multicast group management

  The components of LANE are

- The LANE driver within each end station (e.g., host, server, or LAN access device)
- One or more LANE services (realized via specialized servers) residing in the ATM network

The LANE driver within each end station provides an IEEE 802 MAC-Layer interface that is transparent to higher-layer protocols. Hence, LANE carries IP PDUs across different physical LANs, but in the same IP subnetwork. Within the end station, the LANE driver also translates IEEE 802 MAC-Layer addresses into ATM addresses, using an address resolution service provided by a LANE server. It establishes point-to-point ATM SVC connections to other LANE drivers and delivers data to other LANE end stations. LANE drivers are also supported on access devices (e.g., routers, hubs, and LAN switches) attached to the ATM internetwork. The access devices differ from end stations on the ATM internetwork in that access devices act as a "proxy" for end stations. As such, they must receive all multicast and broadcast packets destined for end stations located on attached LAN segments.

---

[3]This section is based on Ref. [0], with permission.

LAN emulation services are realized using the LAN emulation server (LES), the broadcast and unknown server (BUS), and the LAN emulation configuration server (LECS). LANE services can be implemented in an ATM intermediate system; an end station such as a bridge, router, or dedicated workstation; or a PC. They may also be implemented on ATM switches or other ATM-specific devices. LANE services exist as a single centralized service where the LECS, LES, and BUS are implemented on an end station or ATM switch. But they can also be implemented in a distributed manner, with several servers operating in parallel and providing redundancy and error recovery. LANE services can operate on one or more LAN emulation clients (LECs). For example, the LECS may reside on one end station, which is also an LEC, while the LES and BUS reside on another end station running LEC code.

This combination of LANE drivers and services transparently supports the operation of existing 802.x LAN applications over the ATM internetwork. By using multiple LANE services, multiple 802 LANs can be emulated on a single physical ATM internetwork. This allows LAN administrators to create VLANs[4] [these are also called emulated LANs (ELANs)].

The advantages provided by LANE compare favorably to those of LAN bridging. LAN bridging technology was developed to support the expansion of LANs. Ethernet bridges are transparent and require minimal configuration. Attached PCs do not require any modifications to operate in a bridged environment, saving much of the administrative cost associated with other internetworking technologies. Both LANE and classical bridging support MAC-Layer connectivity between LAN applications. LANE, however, removes the limitations of classical bridging, making it a building block of ATM internetworking.

A single ATM network with LANE supports multiple VLANs. Because each VLAN is distinct from the others, broadcast traffic in one VLAN is never seen in any other VLAN. It does not require any filtering or other mechanisms on stations not in that particular VLAN. The LECS allows dynamic configuration capabilities within the ATM internetwork, eliminating the need to define the physical connection between a host computer and the VLAN(s) to which it belongs. This allows a host computer to be moved from one building to another while remaining a member of the same VLAN.

---

[4]A VLAN is a logical association of users sharing a common broadcast domain.

By using existing 802.x frame types and emulating the behavior of 802.x LANs, ATM network adapters appear to end stations and upper-layer protocols to be Ethernet or token ring cards—or both. Any existing protocol that has been defined to operate over Ethernet or token ring LANs can also operate over ATM LANE without modification. In particular, IP traffic is supported.

As discussed in Chap. 3, in 1995, the LAN Emulation Working Group of the ATM Technical Forum passed the *LAN Emulation over ATM* Version 1.0 specification. It defines the LAN emulation user-to-network interface (LUNI), over which existing LAN protocols operate. The LUNI describes how an end station communicates with the ATM internetwork. The LUNI defines initialization and registration. The LECS (server) controls the assignment of individual LECs (clients) to VLANs, using information contained in the LECS's database as well as information provided by each LEC. The definition of the LUNI model allows independent vendors to implement LANE end stations, while providing interoperability between their products. The LUNI defines initialization, registration, address resolution, and data transfer procedures for the interaction of the LEC and the LANE services. As the ATM Forum continues to develop LANE standards, the current specification allows for a number of implementations in the host computer.

Further work by the ATM Forum will focus on the LAN emulation network-to-network interface (LENNI). Pre-LENNI solutions are possible today and are available from a single vendor or multiple vendors who use the same signaling technique for network-to-network interfaces. The Version 2.0 specification, providing enhancements in the area of redundant servers and Subnetwork Access Protocol/Logical Link Control (SNAP/LLC) encapsulation (per RFC 1483), was nearing completion at press time. LANE Version 2.0 begins to distinguish the elements within the LANE service cloud. It aims at accommodating multiple LES/BUS pairs by defining protocols between them. These protocols provide a level of scalability for LANE, and will support server function redundancy for improved robustness. There are also extensions related to QoS; QoS is designed to manage integrated voice, video, and data traffic in an ATM network. Through the use of different virtual connections, QoS supports applications that require constant, variable, available, and unspecified bandwidth. ATM switches can build a virtual circuit for each application and use QoS information to set up traffic priorities, choose network routes, and manage trunk availability.

Many observers, however, believe that LANE will be difficult to implement as a network discipline across an enterprise because it imposes a large broadcast domain (a single IP subnetwork). Otherwise, if multiple ELANs are utilized, routers are needed to interconnect them; here routers can become bottlenecks. Hence this approach does little to diminish the use of routers in the network or push them to the edges.

7.3.2 Classical IP over ATM

Classical IP over ATM (CIOA) predates LANE and is the method of running LAN traffic over ATM that was developed by the IETF. The IETF's specification is defined to provide native IP support over ATM and is documented in the following RFCs:

- RFC 1483, *Multiprotocol Encapsulation over ATM Adaptation Layer 5*
- RFC 1577, *Classical IP and ARP over ATM*
- RFC 1755, *ATM Signaling Support for IP over ATM*
- RFC 2022, *Multicast Address Resolution (MARS) Protocol*

These protocols are designed to treat ATM as virtual "wire" with the property of being connection-oriented, therefore, as with LANE, requiring unique means for address resolution and broadcast support.

In the CIOA model,[5] the ATM fabric interconnecting a group of hosts is considered a network, called nonbroadcast multiple access (NBMA). An NBMA network is made up of a switched service like ATM or frame relay with a large number of end stations that cannot directly broadcast messages to each other. While on the NBMA network there may be one OSI Layer 2 network, it is subdivided into several logical IP subnetworks (LIS) that can be traversed only via routers.

One of the design philosophies behind CIOA is that network administrators started out building networks using the same techniques that are used today, that is, dividing hosts into physical groups, called subnetworks, according to administrative workgroup domains.[6] Then the subnetworks were interconnected to other subnetworks via IP routers. An LIS in CIOA is made up of a collection of ATM-attached hosts and ATM-attached IP routers that are part of a common IP subnetwork. Policy administration,

---

[5]This section is based on Ref. [101].

[6]As was discussed earlier, however, this need not be obligatory going forward.

such as security, access controls, routing, and filtering, will still remain a function of routers because the ATM network is just "smart" wire.

In CIOA, as in LANE, the functionality of address resolution is provided with the help of special-purpose server processes that are typically colocated. This is accomplished via software upgrades on legacy routers. Each CIOA LIS has an ARP (Address Resolution Protocol) server that maintains IP-address-to-ATM-address mappings. All members of the LIS register with the ARP server, and subsequently all ARP requests from members of the LIS are handled by the ARP server. This mechanism is a little more straightforward than LANE, since for ARP there is only one server, and this server maintains direct IP-to-ATM address mappings.

In the CIOA model, IP ARP requests are forwarded from hosts directly to the LIS ARP server using MAC/ATM address mappings that are acquired at CIOA registration. The ARP server, which is running on an ATM-attached router, replies with an ATM address. When the ARP request originator receives the reply with the ATM address, it can then issue a call setup message and directly establish communication with the desired destination.

One of the limitations of this approach is that CIOA has no understanding of QoS. CIOA has the drawback of supporting only IP because the ARP server is knowledgeable only about IP. In addition, this approach does little to reduce the use of routers, although it does have the effect of separating to a degree the data forwarding function from the IP PDU processing function; in effect, IP PDUs do not have to be examined at the end of each hop, but can be examined at the end of a virtual channel (VC) or path (VP), which may consist of several hops—the challenge is how to identify (address) the VC in question to reach a specific remote IP peer; hence the address resolution function. The CIOA model's simplicity reduces the amount of broadcast traffic and interactions with various servers.[7] In addition, once the address has been resolved, there is a potential that the subsequent data transfer rate may be reduced. However, the reduction in complexity does come with a reduction in functionality.

As is the case with LANE, communication between LISs must be made via ATM-attached routers that are members of more than one LIS. One physical ATM network can logically be considered several logical IP subnetworks, but the interconnection across IP subnetworks, from the

---

[7]By reducing communication that would be required in LANE with the LECS, LES, and BUS, the time required for address resolution can be reduced.

host perspective, is accomplished via another router. Using an ATM-attached router as the path between subnetworks prevents ATM-attached end stations in different subnetworks from creating direct virtual circuits with one another. This restriction has the potential to degrade throughput and increase latency. There are also questions about the reliability of the IP ARP server in that the current version of the specification has no provisions for redundancy: If the ARP server were to fail, all hosts on the LIS would be unable to use the ARP. Finally, CIOA suffers from the drawback that each host needs to be manually configured with the ATM address of the ARP server, as opposed to the dynamic discovery allowed in LANE.

Data transfer is done by creating a VC between hosts, then using LLC/SNAP encapsulation of data that have been segmented by AAL5. Mapping IP packets onto ATM cells using LLC/SNAP is specified in RFC 1483, *Multiprotocol Encapsulation over ATM*. RFC 1483 specifies how data are formatted prior to segmentation. (The RFC documents several different methods; however, the vast majority of host/router implementations use the LLC/SNAP encapsulation. LLC/SNMP specifies that each datagram is prefaced with a bit pattern that the receiver can use to determine the protocol type of the source.) The advantages provided by the encapsulation method specified in RFC 1483 are that it treats ATM as a Data Link Layer that supports a large maximum transfer unit (MTU) and that it can operate in either a bridge or a multiplexed mode. Because the network is not emulating an Ethernet or token ring, like LANE, the MTU has been specified to be as large as 9180 bytes. The large MTU can improve the performance of hosts attached directly to the ATM network.

RFC 1577 specifies two major modifications to traditional connectionless ARP. The first modification is the creation of the *ATMARP message* used to request addresses. The second modification is the *InATMARP message*, which inverts address registration. When a client wishes to initialize itself on an LIS, it establishes a switched virtual circuit to the CIOA ARP server. Once the circuit has been established, the server contains the ATM address extracted from the call setup message calling party field of the client.

The server can now transmit an InATMARP request in an attempt to determine the IP address of the client that has just created the virtual circuit. The client responds to the InATMARP request with its IP address, and the server uses this information to build its ATMARP table cache. The ARP table in the server will contain a listing for the IP-to-

ATM pair for each host that has registered and periodically refreshed its entry to prevent it from timing out. The ATMARP server cache answers subsequent ATMARP requests for a client's IP address. Clients wishing to resolve addresses generate ATMARP messages, which are sent to their server, and locally cache the reply. Client cache table entries expire and must be renewed every 15 min. Server entries for attached hosts time-out after 20 min.

As noted, multicast support is of interest. CIOA provides multicast support via the multicast address resolution server (MARS). The MARS model is similar to a client/server design because it operates by requiring a multicast server to keep membership lists of multicast clients that have joined a multicast group. A client is assigned to a multicast server by a network administrator at configuration time. In the MARS model, a MARS system, along with its associated clients, is called a cluster. The MARS approach uses an address resolution server to map an IP multicast address from the cluster onto a set of ATM endpoint addresses of the multicast group members.

The three primary components of a MARS-based IP over ATM network are

- A top-level server(s) called the MARS
- Zero or more multicast servers that provide second-level multicast distribution
- Clients that utilize IP multicast by building point-to-multipoint paths based on information learned by MARS

Every MARS[8] has at least one client and server contained within a CIOA logical IP subnetwork. To operate the system, clients use the MARS as a means of determining what other hosts are members of a multicast group. In a MARS network there are two modes of operation: (1) full mesh and (2) multicast server. In the full mesh mode, client queries are sent to the server to identify which hosts have registered as members of a class D tree.[9] Next, the client establishes a point-to-multipoint virtual circuit to those leaves. In the second mode, the multicast server acts as the focal point of all multicast packets originated anywhere in the multicast tree. In this case, in order to simulate IP multicast over an ATM network, the multicast server simply retransmits,

---

[8]Typically, the MARS is colocated with the CIOA ARP server.

[9]A class D address is part of the global IP multicast range.

over the ATM multicast connections, all PDUs sent to the IP multicast group by the clients. Because the set of hosts in a multicast group is constantly changing, the MARS is also responsible for dynamically updating the set of clients with new membership information as changes occur along with adding clients to and removing them from the active members.

When running multicast over an ATM network, selecting between the two modes of operation described above is left to the discretion of the network designer. An additional design determination with MARS can be made by adding multiple layers of hierarchy to the distribution tree. For example, multicast clusters may contain the second level of the hierarchy by elevating a client to the roll of a multicast server.

One of the tradeoffs that is conceded in order to gain the simplicity that MARS offers is the required "out of band" control messages used to maintain multicast group membership. In order for clients and multicast servers to send and receive control and membership information, the MARS protocol specifies the setup of a partial mesh of virtual circuits. The MARS maintains its own point-to-multipoint circuits, called the ClusterControlVC, for the members within the cluster.

The ClusterControlVC carries leaf node update information to the clients as members leave and join the multicast session. Each client in a multicast cluster maintains a point-to-point VC to the MARS that is used to initialize itself and for path group change messages. Finally, the MARS manages the multicast servers through point-to-point virtual circuits between each multicast server and the MARS, and through point-to-multipoint circuits, called ServerControlVC, from the MARS to the multicast servers. These circuits are used, like the ClusterControlVC, to pass information from the MARS to keep the cluster membership updated.

The MARS protocol utilizes a set of control messages (see Table 7.5) that are exchanged between the MARS and the clients in order to maintain the group memberships. In addition, the MARS has a special set of messages (Table 7.6) that it exchanges with the second-tier multicast servers (should one exist).

The operation by clients wishing to exchange IP multicast traffic on a MARS system is as follows:

1. Register with the MARS (transmit a MARS_JOIN with only the client's ATM address).

TABLE 7.5

MARS Messages
(to Clients)

MARS_JOIN/LEAVE	Used by clients to register with the MARS as cluster members, and to join or leave particular IP multicast groups.
MARS_REQUEST	Issued by clients when they first join a MARS. The message is sent to the MARS to determine the set of hosts that are part of a multicast group.
MARS_REDIRECT_MAP	Acts as a "heartbeat" mechanism to provide a steady sanity check of the MARS system. The messages are periodically transmitted from the MARS to the clients over the Cluster-ControlVC.
MARS_MIGRATE	Used to force the deletion and subsequent re-creation of a client's point-to-multipoint circuits for a multicast group and replace them with new circuits.

TABLE 7.6

MARS Messages (to
Multicast Servers)

MARS_MSERV/UNSERV	Exchanged between a multicast server and the MARS to signify the multicast server's ability to begin/end being a multicast server for a particular class D multicast session.
MARS_SJOIN/SLEAVE	Propagated on the ServerControlVC by the MARS when it is necessary to inform the multicast servers that their point-to-multipoint shared forwarding VCs may need updating.
MARS_REQUEST	Issued by multicast servers when they are first registering with the MARS to determine the set of hosts that are part of a multicast group.
MARS_REDIRECT_MAP	Determines the ATM address of a second MARS to use in the event of failure.

2. Multicast group join (transmit a MARS_JOIN containing one or more class D IP multicast addresses and the client's ATM address; the MARS then propagates the MARS_JOIN to all clients that have already joined that multicast group, and each sender that receives the MARS_JOIN adds the new client to its ATM multicast connection).

3. Data transmission (transmit a MARS_REQUEST to the MARS containing the class D IP multicast address and the ATM address of the client; the MARS returns a list of ATM addresses associated with the IP multicast address in a MARS_MULTI or, if there are no receivers in the group, responds with a MARS_NAK).

The reader may refer to Ref. [101] for a more extensive treatment of this topic.

7.3.3 MPOA

MPOA can be viewed as solving the problems of establishing connections between pairs of hosts that cross administrative domains (i.e., IP subnets) and enabling applications to make use of a network's ability to provide guaranteed QoS [99][100]. For some time already, manufacturers have released products that separate switching from routing, and allow applications to designate their required QoS. MPOA supports intersubnet cut-through in enterprise networks. The MPOA working group of the ATM Forum is charged with developing a standard approach to forwarding Layer 3 protocols, such as IP or Novell's IPX, transparently over ATM backbones. Building upon LANE, MPOA allows ATM backbones to support legacy Layer 3 protocols and their applications. MPOA will also allow newer Layer 3 protocols and their applications, such as packetized video applications using IP's RSVP, to take advantage of ATM's QoS features over the same ATM backbone. With MPOA, end systems (user clients and corporate servers) can all be just one hop away. In effect, the routing is relegated to the edge of the network. This unbundles the data forwarding function from the IP PDU processing function. Table 7.7 compares two available schemes for routing over ATM [rb]. Also see Fig. 7.4.

MPOA enables the separation of the route calculation function from the actual Layer 3 forwarding function. This provides three key benefits: integration of intelligent VLANs, cost-effective edge devices, and an evolutionary path for clients from LANE to MPOA. In the MPOA architecture, routers retain all of their traditional functions so that they can be the default forwarder and continue to forward short flows as they do today. Routers also become what are commonly called MPOA or route servers and supply all the Layer 3 forwarding information used by MPOA clients, which include ATM edge devices as well as ATM-attached hosts. Ultimately, this allows these MPOA clients to set up direct "cut-through" ATM connections between VLANs to forward long flows without having to always experience an extra router hop [99][100].

A design desideratum in MPOA is to ensure that both bridging and routing are preserved for legacy LANs and the VLAN topology in use. An MPOA network uses LANE for the bridging function. As implied in the LANE discussion, an emulated LAN's scope (that is, an ELAN) is a single Layer 3 subnet,[10] whereas MPOA is focused on intersubnet (IP sub-

---

[10]As noted in the previous footnote, this implies that routers would be needed to connect ELANs.

TABLE 7.7

Schemes for Routing over ATM

1. RFC 1483 multiprotocol encapsulation

   ■ Simple.

   ■ ATM VCs are defined between every pair of routers.

   ■ Limitation: Full connectivity leads to the *N*-squared problem: As routers are added to the network, routing tables grow quadratically, since every router needs a pointer to every other router, and routing updates consume increasing amounts of bandwidth.

2. ATM Forum's MPOA

   ■ Designed primarily for LANs/campuses, it replaces a collapsed backbone with a "distributed" or virtual router.

   ■ LAN switches and other edge devices become the virtual router's I/O ports, the route server is the central processor, and ATM switches are the backplane.

   ■ Workstations and servers belong to virtual subnetworks.

   ■ PDUs with destinations within virtual subnetworks are bridged at Layer 2 using LANE.

   ■ PDUs with destinations in other virtual subnetworks are sent to the route server, which forwards them to the destination device. Simultaneously, the route server downloads Layer 3 information to the source device, and NHRP determines the ATM address of the destination. Subsequent PDUs between the same source and destination cut through the ATM backbone directly, bypassing the route server.

net) connectivity. Using LANE within the MPOA specification provides a number of benefits to the user, including the fact that it allows backward compatibility, as an MPOA network can be built with MPOA clients as well as LANE clients. In fact, the default operation for an edge device can be LANE until it learns more, simplifying the topology configuration and startup operation.

For the Layer 3 forwarding function, MPOA is adopting and extending the Next Hop Routing Protocol (NHRP). NHRP is designed to operate with current Layer 3 routing protocols and thus does not require any replacement for or changes to those protocols. In order to set up a direct ATM connection between two ATM-attached hosts or between an ATM-attached host and an edge device, the ATM address of the exit point that corresponds to the Layer 3 address of the desired destination must be determined. An ATM-attached host can send an

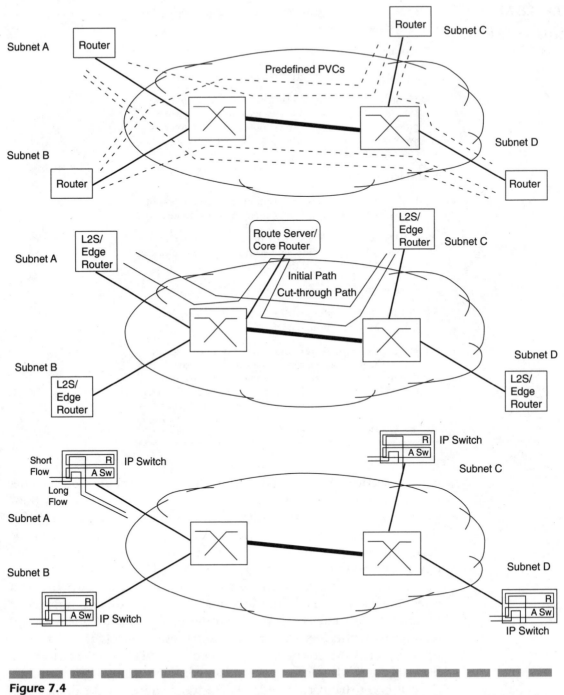

Figure 7.4
IP over ATM technologies. *Top:* multiprotocol encapsulation over ATM. *Middle:* MPOA. *Bottom:* IP Switching.

NHRP query to an MPOA server, which has been getting reachability information from routing protocols such as Open Shortest Path First (OSPF). The MPOA server may then respond with the ATM address of the exit point or ATM-attached host used to reach the destination Layer 3 address, or it may forward the query to other MPOA servers if it does not know the answer. Ultimately, the MPOA server that is serving the client that can reach the destination Layer 3 address will know the answer and reply. Once the reply arrives at the source, a direct "cut-through" ATM connection can be set up [99][100].

In a VLAN environment, NHRP will reply with the ATM address of a router that serves the destination's VLAN. If this VLAN has more than one router connected to it, there is no guarantee that the reply will address the router closest to the destination. When it does not, then that router must bridge the data packets to the closest router, which will then forward them on to the destination. MPOA has defined mechanisms in addition to NHRP that allow the MPOA servers to give out Layer 3 forwarding information to edge devices that represent the optimal exit point for a given destination so that this potential for excess hops is avoided.

MPOA is already a standard as of press time, whereas the IP Switching technologies are just being standardized. MPOA is a partial step toward network cost reduction as related to the deployment of routers; this is accomplished by sharing Layer 3 intelligence (route management in the route server) among edge devices. From an economic point of view, developers realize that MPOA components must be priced so that a local MPOA implementation is cheaper than a large backbone router and multiple LAN switches. There already was available MPOA equipment by press time, notably Newbridge's Vivid. The Vivid kit includes a workgroup switch, a Yellow Ridge Ethernet switch with an ATM uplink, a route server, and system manager software. This Vivid equipment provides an alternative to both the backbone router and the workgroup switches.

In an Internet application, MPOA could theoretically be deployed to support cut-through, as follows:

■ Install an MPOA system in place of a router in a central office (CO).

■ Install edge devices (e.g., the LAN switches of Newbridge's Vivid product family) that connect via LANs to remote access servers (which support modem pooling for user access).

■ Implement the virtual router out across COs over the existing ATM infrastructure.

■ Connect Web servers to remote edge devices in remote COs.

■ Utilize MPOA to build virtual private networks (VPNs).

MPOA has some limitations: To ensure both availability and high performance, route servers may possibly be needed in nearly every CO. Some claim that dependence on a server could limit scalability, although this is debatable since the growth of 800/888 service has not been limited in any way, in spite of the fact that it also uses centralized telephony route servers. Overall, MPOA is more directly applicable to campus networks than to large-scale networks such as Internet backbones.

7.3.4   Network Layer Switching

Layer 3 switching, also known as IP Switching, in all its forms, has two goals: (1) to find a way for internetworks (especially the Internet) to scale economically, and (2) to bring effective QoS support to IP. The following view was typical at press time: "While many carriers have committed to ATM for Layer 2 transport, the question of how best to route IP traffic remains an open question. How many carriers will bet on new MPOA protocols to keep their route servers in sync when time-tested solutions like OSPF aren't broken?" [rb]. Some approach the scaling challenge with faster routers and higher-speed lines. Several vendors have launched super-router initiatives and have announced router-over-SONET interfaces; many, however, believe that solutions based on switched technologies like ATM have much more potential. Switching can replace latency-prone, processing-intensive Layer 3 "hops" with more efficient Layer 2 connections.

Various Network Layer Proposals Advanced   The approaches to IP support over ATM covered thus far evolved in the mid-1990s. The late 1990s have seen the emergence of various schemes to address several limitations of traditional IP routing implied by the discussion in Secs. 7.1 and 7.2, including (1) the need for meshed or near-meshed physical networks, (2) the requirement that Layer 3 processing be performed at the endpoints of each link (in effect collapsing data forwarding/transmission, IP processing, and topology discovery into a single, obligatory function at each link endpoint), (3) the relative complexity of Layer 3 processing, (4) the duplication of Layer 2 and Layer 3 functionality (e.g., addressing), and (5) the relatively poor use of improved Data Link Layer technologies (e.g., ATM) by IP.[11]

To address these concerns, a number of vendor-specific and standards-based solutions have been proposed and/or are under development or deployment. Notable vendor-specific solutions include Cisco's Net-Flow/Tag Switching technology and Ipsilon's IP Switching technology. Standards-based solutions include MPOA, already discussed, and Multi-protocol Label Switching (MPLS), which is the evolving specification for cohesive Layer 2/Layer 3 switching/routing.

The ability to switch data based on very fast hardware table lookups on the MAC or ATM addresses leads to very fast and reliable networks. However, these technologies also pose problems of scalability and complexity that are seen by large ISPs or in large enterprise networks. High degrees of scalability can be difficult to achieve with a Layer 2 switched network because the address space is nonhierarchical. Combinations are difficult to find with a pure Layer 2 switched network, so network managers have traditionally used Network Layer protocols, like IP, to fill that void. Of the problems introduced by large Layer 2 networks, some of the more pressing concerns have to do with smoothly integrating Layer 2 switching with Layer 3 switching. A number of solutions were emerging as of press time.

The goals of Network Layer switching are to provide new means for interworking Layer 2 and Layer 3 technologies. Where the interworking differs from previous protocols is that the functionality of traditional Network Layer protocols, such as IP, plays a more important role in the overall control of the network. In a Network Layer switched environment, all of the ATM switches understand and are capable of routing IP packets using protocols like the Border Gateway Protocol (BGP) or OSPF. In this model, the benefits of ATM that are applied to network design are basically speed and traffic control.

Recent experience with ATM and IP integration has given network designers a perspective on the technology that deviates somewhat from ATM's initial goals. Most first-generation ISP or enterprise ATM deployments utilized ATM for either its speed or its ability to provide strict controls over traffic flow. The high speeds come in the form of access facilities (and networks) that operate at 155-Mbps or 622-Mbps optical interfaces, and traffic control is achieved via PVC assignment. While this

---

[11]This section is based on Ref. [dmsw].

use of ATM has been useful, it is a very coarse method for designing a network that can sometimes lead to unexpected traffic flows. Standards activity and recent network deployments have focused attention on issues surrounding traffic engineering and the ability to use the Layer 2 network to explicitly define the route that data follow.

The body of work in the field of Network Layer switching can be subdivided into two categories based on the level of granularity that is applied when mapping IP traffic to ATM virtual circuits. These models are called *flow*-based and *topology*-based. From a high-level view of network switching, one can think of the flow-based models as building a network out of router/switching devices in which unique ATM virtual circuits are created for each IP *conversation,* where a conversation is synonymous with a file transfer or WWW session. In the topology proposals, the router/switching devices use their ATM fabrics to create ATM virtual circuits that can carry all of the traffic between pairs of subnetworks or on IP routes.

Regardless of the exact technical approach, the goals of Network Layer switching are similar because Network Layer switching's fundamental motivations are to remove excess computational processing done during PDU transmission by clearly dividing routing from forwarding, then to remove routing from the process whenever possible. In both the flow-based and topology-based models, the ATM switch must be aware of IP and capable of participating in IP routing protocols. However, a clear division between the processes of forwarding and routing is still maintained. Once the control process (i.e., routing) has detected either a route or a flow, it removes itself from the communication path and employs high-speed forwarding from the ATM fabric.

Cisco's NetFlow and Tag Switching Technology   Cisco has been the first major player to address the issue of Layer 3 cut-through via its Net-Flow/Tag Switching technology. This technology supports flow-oriented switching for multiple protocols. The approach is to "learn once—switch many times." Cisco has positioned Tag Switching as a LAN technology and NetFlow as a WAN technology—fundamentally they are similar in concept.

Tag Switching is Cisco's answer to Ipsilon's IP Switching, and it is positioned for networks with 30 to 40 backbone routers. Tag Switching utilizes traditional routing protocols to identify and maintain paths. The novelty is that all paths leading to the same destination (e.g., an IP subnet) are assigned the same "tag." The Tag Distribution Protocol (TDP) maintains tables in each node that relate tags to destinations. As PDUs enter the network, a router handles Layer 3 processing and assigns each one a tag; at subsequent nodes, which can be router, ATM switches, or frame relay switches, Layer 3 processing is bypassed, and PDUs are forwarded based only on their tags. Routing hops are eliminated from the interior of the network [rb]. Hence, this approach enables a form of cut-through, in that the routing decisions are ostensibly relegated to the edge of the network.

A flow is a unidirectional sequence of packets between a given source and destination. Issues related to flows are: "What do I use to define a NetFlow?" "What determines the start of a NetFlow?" "What determines the end of a NetFlow?" "How does one time-out a NetFlow entry?" Net-Flow granularity can be defined in terms of application (Application-Layer applications such as Telnet, FTP, etc.), Transport Layer protocols (e.g., TCP, UDP), Network Layer IP parameters (e.g., IP address), and Data Link Layer protocols (e.g., Ethernet, token ring).

The IP header contains a protocol field (the 10th byte) that can be used to define a flow (e.g., the protocol could be TCP, UDP, etc.). In turn, the UDP and TCP headers contain port numbers that define the nature of the data being carried (e.g., port 53 for DNS, port 161 for SNMP, port 23 for Telnet, port 21 for FTP, port 80 for HTTP). Hence, NetFlow granularity can be defined at the TCP/UDP source or destination port, IP type, and IP source or destination address. The NetFlow flow can start with a TCP SYN flag and terminate with a TCP FIN flag.

A router has various kinds of memory, specifically packet memory and system memory. With NetFlow enhancements to the router, a Net-Flow cache is allocated. An incoming frame is first copied to packet memory. Normally, IP processing takes place to determine the route; i.e., the destination address is removed from the PDU and the routing table is consulted, so that the PDU can be sent to the exit interface. In a NetFlow-enabled router, when a new frame arrives, there may be no match in the NetFlow switching cache. Hence, the PDU is copied to the system buffer for processing. A lookup in the Layer 3 network address table is undertaken to see where the frame should be routed. The Net-Flow switch cache is initialized, and the frame is sent to the exit interface. When the next frame is copied to packet memory, a match is found in the NetFlow switching cache. This implies that the frame can now be sent directly to the output interface without having to go through the additional IP routing processing, specifically frame de-enveloping and routing table lookup, which can be (relatively) demanding in terms of resources.

Tag Switching is applicable at the campus network level, whereas NetFlow is positioned at the WAN level. Tag Switching addresses the throughput, scaling, and traffic engineering issues of corporate enterprise networks. It permits, according to the vendor, a graceful evolution of routing, and is intended to allow integration of ATM and IP. Tag Switching combines Layer 3 routing with label-swapping forwarding (such as that available on Layer 2 ATM/frame relay networks). The simplicity of Layer 2 forwarding offers high performance; a separation of forwarding for long flows and routing aids the evolution of routing. Refer again to Fig. 7.4.

Forwarding is based on a label-swapping mechanism, as well as a control component that is used to maintain and distribute bindings. The router maintains a Tag Forwarding Information Base (TFIB), whose entries include the incoming tag and one or more subentries such as outgoing tag, outgoing interface, and outgoing MAC address. TFIB is indexed by the incoming tag; TFIB may be per box or per incoming interface. The forwarding algorithm works as follows: (1) Extract the tag from the incoming frame, (2) find the TFIB entry with the incoming tag equal to the tag on the frame, (3) replace the tag in the frame with the outgoing tags, and (4) send the frame to the outgoing interface. In working this way, the label-swapping mechanism is really like an ATM switch. Note that the forwarding algorithm is Network Layer independent. The TPD is used to distribute tag bindings to neighbors; the protocol sends information only if there is a change in the routing table and the device does not have a label.

Because tags correspond to destinations rather than source-destination pairs or traffic flows, tag population grows at order N rather than N-squared; this makes scaling feasible in large enterprise networks and the Internet. Tags are allocated in advance; therefore, there are no performance penalties to short-lived flows or to the first PDU of long-lived flows.

PDUs carry the tag between the IP datagram and its Layer 2 envelope. (With IP Version 6, tags will be included in the Layer 3 flow label field.) Each router examines the tag and sends the PDU directly to the output port, bypassing normal routing processing. This implies that Tag Switching can be used without having to immediately convert the physical network to ATM. Performance improvements of 10 to 20 percent are possible. The use of ATM, however, supports broadband. Most enterprise networks today can no longer be designed with T1 (1.544-Mbps) links, especially in the presence of switched 10-Mbps Ethernet, 100-Mbps Ethernet, and Gigabit Ethernet. Hence, a conversion to Tag Switching, first without an ATM infrastructure and then with an ATM infrastructure, can be considered a reasonable migration strategy. In an ATM-based backbone network, VCIs are utilized as tags. TDP sets up the correspondence between routes and VCI tags; no ATM Q2931 signaling is required. Routers at the edge of the network examine the incoming PDU's IP destination address and assign it the proper VCI. ATM switches support the forwarding function, and do so in an effective manner. The internetwork itself remains connectionless, since there are no end-to-end (end-system-to-end-system) virtual circuits, and switches route around network failures. However, the issue of ATM interworking is not fully solved by Tag Switching, because PDUs from different sources destined for a specific destination end up sharing a VCI, and cells may get interleaved, which is a problem in ATM. MPLS aims at addressing this issue.

IP Switching   IP Switching is a proprietary (not sanctioned by the ATM Forum or IETF) networking technology advanced by Ipsilon Networks that combines the control of IP routing with ATM speed, scalability, and Quality of Service to deliver millions of IP packets per second throughput to intranet and Internet environments.[12] Ipsilon wanted to "make IP go fast, but also to create a complete paradigm shift." IP Switching implementations have already propagated from network interface cards to edge systems, telecommunications devices, and backbone, campus, and workgroup switches.

An IP Switch implements the IP stack directly onto ATM hardware, allowing the ATM switch fabric to operate as a high-performance Data Link Layer accelerator for IP routing. An IP Switch delivers ATM at wire speeds while maintaining compatibility with existing IP networks, applications, and network management tools.

Using intelligent IP Switching software, an IP Switch dynamically shifts between store-and-forward routing and cut-through switching based on the needs of the IP traffic, or flows. An IP Switch automatically chooses cut-through switching for flows of longer duration, such as File Transfer Protocol (FTP) data, Telnet data, Hypertext Transmission Protocol (HTTP) data, and multimedia audio and video. It reserves hop-by-hop store-and-forward routing for short-lived traffic, such as Domain Name Server (DNS) queries, Simple Mail Transfer Protocol (SMTP) data, and SNMP queries. The majority of data are switched directly by the ATM hardware, without additional IP router processing, achieving millions of PPS throughput. See Fig. 7.4.

---

[12]This section is based on Ref. [i].

One of the advantages that IP Switching offers is the interoperable network architecture that has resulted from the multivendor acceptance of the technology. While router improvements continue to be limited to individual platforms, IP Switching can become a networkwide solution. According to the vendor, IP Switching with its published protocols provides cooperative benefits to all peer participants: direct cut-through connections across the network, and low latency compared to traditional router networks.

In a classical internetwork, every router is connected to just a few neighbors, in order to keep routing tables to a manageable level from a size and access speed perspective. Routing protocols (see below) are used to determine paths between endpoints and recover automatically when a node or trunk fails. However, the problem with this approach is that a path may entail many Layer 3 hops, producing unacceptable delay and unpredictable performance.

To support effective communication, the exchange of appropriate routing and status information among routers is required. The routers exchange information about the state of the network's links and interfaces and about available paths, based on different metrics. Metrics used to calculate optimal paths through the network include cost, bandwidth, distance, delay, load, congestion, security, QoS, and reliability. Routing protocols are used as the means for exchanging this vital information. There are two routing protocol types for status dissemination: (1) protocols that operate within an autonomous network,[13,14] called interior gateway[15] protocols (IGPs), and (2) protocols that operate between autonomous networks, or exterior gateway protocols (EGPs). While any suitable protocol may be (independently) used for route discovery, propagation, and validation, within an autonomous network, autonomous networks must make routing information available to other autonomous networks using a systemwide protocol. So, within an autonomous network, an interior routing protocol is used; and between autonomous networks, an exterior routing protocol is used. Key IGPs include

- Routing Information Protocol (RIP)
- Open Shortest Path First (OSPF)

---

[13]An autonomous network is typically an administrative domain.

[14]In order to modernize the parlance, we refer to "autonomous networks" or "autonomous subnetworks" rather than "autonomous systems."

[15]The use of the term *gateway* to refer to a router is based on traditional Internet nomenclature.

- Cisco's Interior Gateway Routing Protocol (IGRP)
- Intermediate-System to Intermediate-System (IS-IS) Protocol

Key EGPs include

- Exterior Gateway Protocol (EGP)
- Border Gateway Protocol

The three protocols commonly used in the TCP/IP context are RIP, IGRP, and OSPF.

Two methodologies are used for information dissemination: *distance vector* and *link-state*. Routers that employ distance vector techniques create a network map by exchanging information in a periodic and progressive sequence. Each router maintains a table of relative costs (hop count or other weights, such as bandwidth availability) from itself to each destination. The information exchanged is used to determine the scope of the network via a series of router hops. After a router has calculated each of its distance vectors, it propagates the information to each of its neighboring routers on a periodic basis—say, once every 60 s.[16] If any changes have occurred in the network, as inferred from these vectors, the receiving router modifies its routing table and propagates it to each of its own neighbors. The process continues until all routers in the network have converged on the new topology. Distance vector routing was the early kind of dynamic routing. Distance vector protocols include RIP, IGRP, and DECnet Phase IV. Distance vector protocols can be implemented in a reasonably simple manner. However, they suffer from a number of limitations, including

- Periodic broadcast of routing information, using up network bandwidth
- Susceptibility to routing loops
- Slow convergence, particularly in large enterprise networks

Enhancements to the distance vector method have been developed to speed convergence and reduce the chance of routing loops. These enhancements are *holddown, split horizon,* and *poison reverse.* Refer to Ref. [99] for more detailed information on these topics.

---

[16]More pedantically, periodic updates are not a property of distance vector routing, they are the mechanisms of choice (over unreliable links). Broadcast is used to provide reliability through retransmission.

IP Switching utilizes the same topology and the same routing protocols as conventional routers, but replaces Layer 3 hops with Layer 2 switching. The N-squared problem of multiprotocol encapsulation (RFC 1483) goes away, and network performance improves. Each Ipsilon IP Switch comprises a router (called IP Switch Controller) and an ATM switch. The router exchanges topology information with other IP Switches and provides Layer 3 store-and-forward services, while the ATM switch forwards cells at broadband speed. The software can recognize flows (a flow being a coherent stream of PDUs between the same source and destination). The IP Switch analyzes each flow and classifies it as short- or long-lived.[17] Short-lived flows (e.g., SNMP queries, WWW URLs, DNS packets) are routed by traditional IP-level methods over default VCs and incur normal router latency. Longer flows like file transfers are assigned separate VCs that bypass the IP-level routing processing, and so can be forwarded at much higher speeds [rb].

In an IP Switched network, all hosts and switches communicate through a common set of cooperative protocols—the Ipsilon Flow Management Protocol (IFMP, IETF RFC 1953, 1954) and the General Switch Management Protocol (GSMP, IETF RFC 1987)—to optimize short- and long-lived conversations between a sender and receiver. Routing decisions need be made only once. As soon as longer-lasting flow data have been identified and cut through, there is no need to reassemble their ATM cells into IP packets at intermediate switch points. Thus, traffic incurs minimal latency, and throughput remains optimized throughout the IP Switched network. End-to-end cooperation also enables QoS implementation since, with cut-through switching, policies naturally span administrative boundaries. Hence, flows can be utilized to support flow-binded QoS: by analyzing IP headers, the IP Switch can relate individual flows to performance requirements and request ATM VCs with the proper type of service. Flows emanating from a time-critical application (if it can be identified in some manner, e.g., port number, IP address, etc.) might receive highest priority, whereas ordinary file transfers would run at low priority [rb].

Multivendor support for IP Switching can result in a new economic model for high-performance networks. Customers can select the best platform according to their particular needs in a range of different network environments—from the campus, to the Internet, to the carrier

---

[17]Ipsilon estimates that longer flows constitute more than 80 percent of internetwork traffic.

networks, and even at home. They can choose from an array of IP-based supporting applications, platforms, and tools. And they can work with any number of solution providers, since no single vendor dominates the IP Switching landscape.

Different network elements can implement IP Switching protocols to deliver ATM-accelerated IP while maintaining multivendor, multiplatform interoperability. There is applicability for IP Switching across workgroup, campus backbone, Internet, WAN, and broadband access environments. Early applications included IP Switching in the campus backbone, IP Switching in workgroups, IP Switching in the Internet and across the WAN, and broadband access to IP Switched networks.

One of the limitations of the technology is how it utilizes ATM VCs. Flows are associated with application-to-application conversations, and every long-lived flow gets its individual cut-through VC. This works for relatively small campus networks, but in large network environments, millions of individual flows would quickly exhaust VC tables. Many switches are also limited in the number of VCs they can actually maintain. Furthermore, the constant requests for new VCs can easily overwhelm the switch. A modification may be required in large enterprise networks or on the Internet to support flows with less granularity.

Multiprotocol Label Switching   The Multiprotocol Label Switching (MPLS) Working Group has been charged by the IETF with developing a label-swapping standard for Layer 3 switching. The group started with Cisco's Tag Switching and IBM's nearly identical ARIS (Aggregate Route-based IP Switching). As was noted above, the issue of ATM interworking is not fully solved by Tag Switching, because cell/PDU interleaving occurs when the tag is identified with the VCI. Two schemes to address the issue were being considered by MPLS at press time [rb]:

1. The ATM switch merges multiple VCs into a single VC without interleaving. If two PDUs arrive at the switch simultaneously, the switch buffers cells from one PDU until the other PDU leaves. This approach may require additional hardware in the ATM switch to buffer colliding PDUs. Cisco reportedly plans to use the buffer approach as it adds Tag Switching to the Stratacom BPX ATM switch.

2. The network grows a tree upward from the egress point using ATM VP labels, one VP per egress point. By convention, each source point uses a different VC within the VP. Using VCIs inside the VP, the destination switch can sort out interleaved cells from separate sources.

Although more VCs are used in this method, the amount of state information is still of order N, where N is the number of destinations. This approach needs no new hardware. Ascend reportedly has chosen this approach, and work was underway at press time.

7.4   ATM-like Services without ATM

ATM switches enable carriers to support QoS-based connectivity. Many carriers now offer ATM services. It is true that some carriers cannot truly measure usage (traffic contracts) on their current-generation switches; switch vendors need to develop better tools. It is also true that it is challenging to offer QoS across a network (that is, across several switches) with current technology, since most vendors have focused on delivering QoS over a single switch (see Chap. 10). So, in some cases, carriers are reluctant to offer QoS guarantees to users (e.g., constant bit rate services). However, the interested corporate planner can generally secure ATM services and obtain the needed QoS, although the carrier generally has to overprovision the network to guarantee such QoS.

People looking at the Internet without postulating end-to-end or edge-to-edge ATM are advancing a number of techniques to achieve IP-level QoS. Some of the available techniques are (1) 3Com's Priority Access Control Ethernet (PACE), which supports QoS for video applications on 100-Mbps Ethernet; (2) Cisco's Weighted Fair Queuing for buffer management in routers to reduce delay and jitter; (3) Cisco's Tag Switching/NetFlows; and (4) RSVP.

The Integrated Services Architecture (ISA) approach now under development by the IETF supports QoS in Internet environments, in conjunction with the RSVP signaling protocol. ISA uses a setup protocol whereby hosts and routers signal QoS requests into the network. It defines traffic and QoS characteristics for a flow. Traffic control mechanisms control traffic flows within a host/router to support the required QoS. Elements of the model include [mc]

■ *Classifier,* which maps PDUs to a service class

■ *Packet scheduler,* which forwards packets based on service classes

■ *Admission control,* which determines if the QoS requests can be met

■ *Setup protocol,* which sets up routes (this could be RSVP or ST-2)

ISA will support three QoS classes as follows:

1. *Guaranteed service.* This service provides a guaranteed minimum delay end-to-end. Real-time applications can make use of this service. Leaky bucket, reserved rate, and weighted fair queuing are used for application control. The underlying transport mechanism is ATM [constant bit rate (CBR) or real-time variable bit rate (rt-VBR)].

2. *Controlled load service.* This service provides a best-effort end-to-end capability with a load baseline. Applications that are sensitive to congestion can make use of this service. Leaky bucket methods are used for application control. The underlying transport mechanism is ATM [non-real-time variable bit rate (nrt-VBR) *or* available bit rate (ABR) with a minimum cell rate support].

3. *Best-effort service.* This service provides a best-effort end-to-end capability. Legacy applications can make use of this service. The underlying transport mechanism is ATM [unspecified bit rate (UBR) or ABR with a minimum cell rate support].

ISA requires work for each Layer 2 technology. Hence, the IETF is utilizing different subgroups to look at Ethernet, token ring, and ATM. Utilizing ISA methods, the Internet can be redesigned for real-time applications (e.g., real-time video). However, the overall performance efficiency at the network level remains to be understood (i.e., how many customers can be supported over a given router or link). Some observers now suggest that ATM should support ISA (specifically, there have to be mechanisms to map the ISA classes to the ATM classes).

As noted, RSVP is utilized in ISA. It augments best-effort connectionless services with a QoS request/allocation mechanism; to a degree, this is similar in scope to ATM (see Table 7.8). The RSVP updates the classifier with the Filterspec and the scheduler with the Flowspec. New software and hardware are needed on routers and end systems to support the QoS negotiations. There are expectations that by 1999 there should be relatively widespread deployment of RSVP-based systems, although at press time there still were major design, implementation, engineering, and standardization challenges. It should be noted that with RSVP, the network still utilizes routers and IP.

TABLE 7.8

RSVP/ATM
Comparison

Feature	RSVP	ATM
Initiation	Receiver-driven	Source-driven
Directionality	Unicast/simplex	Duplex
Uniformity	Allows receivers with heterogeneous QoS for a given session	Homogeneous QoS per SVC
QoS renegotiation	Allows dynamic reconfiguration of resources	Requires new setup (new PVC/PVP/SVC) to support a change (except for ABR)
Length of session	Reservations expire (time out)	Permanently reserved for the connection until the connection is dropped
Maturity	Under development at press time	Well developed at press time

7.5 References

[0] Fore Systems, promotional material, used with permission.

[99] D. Minoli and A. Alles, *LAN, ATM, and LAN Emulation Technologies*, Artech House, Norwood, Mass., 1997.

[100] D. Minoli and A. Schmidt, *Client/Server over ATM*, Prentice-Hall/Manning, Greenwich, Conn., 1997.

[101] A. Schmidt and D. Minoli, *MPOA*, Prentice-Hall/Manning, Greenwich, Conn., 1998.

[dmeco] D. Minoli and E. Minoli, *Web Commerce Handbook*, McGraw-Hill, New York, 1998.

[dmsw] D. Minoli and A. Schmidt, *Network Layer Switching Technologies*, John Wiley, New York, 1998.

[i] Ipsilon Networks, promotional material, used with permission. Personal communication with J. Doyle, Ipsilon.

[mc] J. McQuillan, *The NGN Executive Seminar,* March 20, 1997, New York.

[rb] R. B. Bellman, "IP Switching—Which Flavor Works For You?" *BCR,* April 1997, pp. 41–46.

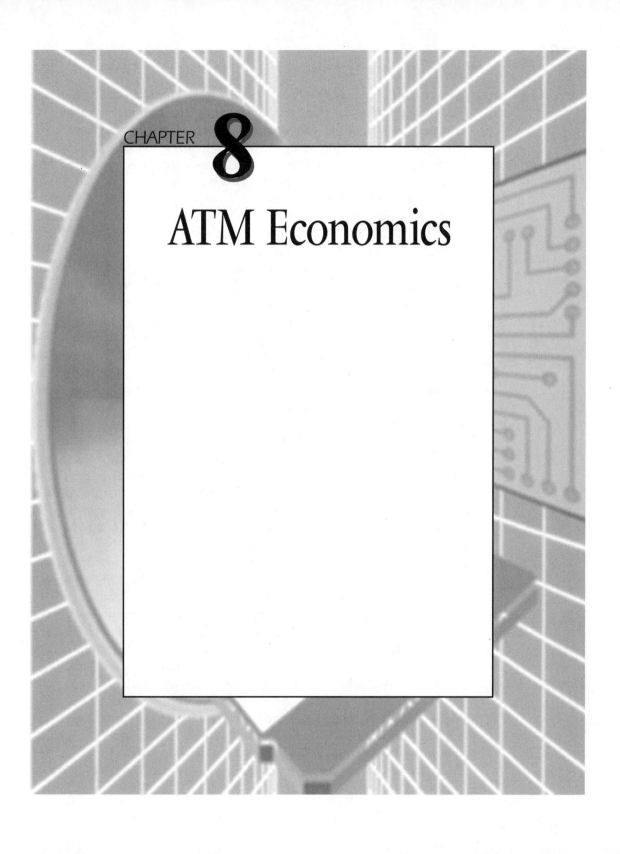

CHAPTER **8**

ATM Economics

ATM technology and services are now being brought to the market at a fairly rapid pace, to support IP and other applications. A variety of equipment elements in support of enterprisewide (data) applications is becoming available, including premises and network-based equipment. Some equipment supporting video and multimedia is also appearing. Over a dozen vendors manufacture public network switches, and several dozen vendors supply a full array of user-level products, e.g., workgroup switches, routers, network interface cards. Some of this equipment has already reached second-generation stage. Over a dozen carriers in the United States have introduced ATM WAN services.

At least for data applications, ATM has reached stability in terms of standards that support a wide range of capabilities and facilitate multivendor interoperability. Work on video and voice is underway, and products are being developed. As noted in Chap. 7, recent emphasis has focused on campus-level ATM, particularly support of IP (e.g., classical IP over ATM), LAN emulation, and improved ways of utilizing ATM for enterprise (routed) networks based on IP and IPX [e.g., Multiprotocol over ATM (MPOA)].

The significant questions now facing prospective users relate, therefore, not to the technology's feasibility and availability, but to ATM's cost-effectiveness, particularly as compared to alternatives. At the local level, there are several alternatives, including switched Ethernet, switched 100-Mbps Ethernet, and Gigabit LANs; appropriate tradeoffs and comparisons are needed here. However, at the wide area level, ATM is the only broadband technology currently available. Furthermore, it should be noted that ATM is used as an internal technology to support native-speed LAN interconnection over WANs, e.g., 10/100-Mbps Ethernet, FDDI, and token ring bridging over ATM (specifically, transparent LAN interconnection services offered by a number of carriers, including TCG); hence, it is also useful to look at what the costs associated with those services are. ATM is also used for Internet backbones.

ATM hardware and services continue to show improvements in terms of costs. For some applications (e.g., WAN and campus backbone connectivity), ATM is indeed very cost-effective; for other applications (e.g., desktop connectivity), ATM has a hard time competing with switched 10 Mbps Ethernet and with 100-Mbps Ethernet.

Planners must examine the following important economic and technology issues discussed in this chapter: cost structures and factors that apply to networking in general and ATM in particular; ATM deployment options; the types of available ATM products and carrier services and their associated costs; and future cost trends for ATM network components [dmdp].

8.1 Decision Approach

Specific corporate decisions regarding ATM can be made only by taking into account the exact environment of the organization in question, in the context of its embedded technology base, network geographic reach and topology, business applications, and corporate networking objectives. Nonetheless, the generic observations presented in this chapter will serve as an input to the necessary analysis. The technology has matured, so that rather than focusing on technology feasibility, as was the case in the mid-1990s, the planner can focus on specific cost elements in comparison with other specific alternatives (when available).

One point worth noting on the part of the planner is that a lot of the trade press coverage in the early to mid-1990s has been focused on what the *developers* have to do to bring this technology to the market, rather than on *end-planner utilization* of this technology once it becomes available. A lot of the debates (e.g., Do we use AAL3/4 or AAL5 for data? Is input buffering better than output buffering? Do we use rate-based or credit-based flow control for available bit rate service? How do we do traffic shaping and tagging, by one or two leaky buckets? How does engineering build traffic management algorithms so that if a carrier chooses to tremendously overbook bandwidth, the grade of service is within specification?) should have been of interest mostly to the technology developers, not to consumers. These issues are "internals" and, to the ultimate consumer of technology, are mostly fixed parameters. Planners should therefore now focus on ATM price/feature considerations, rather than on research-level alternatives to possible implementation of ATM algorithmic mechanisms.

Naturally prices changes over time. However, the information presented herewith, specifically the forecasts included in Sec. 8.5, should continue to be on the mark for several years.

8.2 Cost/Feature Reference Modeling

Ultimately, cost is a key consideration in any business decision. Extensive research undertaken by the authors during the past few years with regard to user views on ATM in general and cost issues in particular strongly supports the statement that user organizations consider cost-effectiveness a critical factor.[1] Cost-effectiveness is important in view of the fact that many information systems (IS) departments are now operating with single-digit budget growth (about 6 percent per year for the period 1995–1999, based on recent large-scale survey information), while at the same time being asked by senior management to undertake IS modernization such as transition to client/server systems, intranets, newer PCs, higher bandwidth, and the introduction of workflow and/or cooperative computing systems, and to support the overall organization's increased reliance on computerization, with ensuing needs for technology deployment, all to improve the global business competitiveness of the organization and to enable electronic commerce.

We also see an increasing desire on the part of senior management of many organizations to benchmark their communications and computing costs, in appropriate productivity units, against their competitors' costs, with the goal of bringing these costs to parity with those of the industry at large. This is motivated by the fact that in most organizations, communications and computing are still viewed as cost centers rather than profit centers. These cost elements typically involve 1 to 12 percent of the gross revenue of a company, with 5 percent being the average. For example, a typical bank with $1 billion of revenue per year would spend on the order of $40 million per year on IS and communications (about 4 percent), and a petrochemical company with $8 billion of revenue per year would spend on the order of $320 million (also about 4 percent).[2] Table 8.1, compiled from various industry sources, provides the typical allocation of expenditures to subelements of the IS operation.

From this table, the $8 billion petrochemical company would spend $120 million on personnel, or have about 1500 IS/communications people at an average of $75,000 per person. The embedded base of IS/communications equipment can be approximated heuristically as being about equal to the yearly expenditure. For example, the $8 billion petrochemical company would have an embedded base of equipment of $320 million.[3]

These numbers can assist the planner in getting a first-order view of the environment in which ATM is to be deployed. For example, if a company with 40,000 employees, all with network-connected PCs, ulti-

---

[1] Other critical factors include network management and interoperability.

[2] As a quick point of comparison, assume that the average worker in a company generated $200,000 worth of revenue; then an $8 billion company would have 8,000,000,000/200,000=40,000 employees. This implies that the IS/computing cost per employee (per seat, as this is sometimes called) would be $8000 per year.

[3] The interested reader may consult the text *Analyzing Outsourcing: Reengineering Information and Communication Systems,* by D. Minoli (McGraw-Hill, 1994), for additional modeling information and for sources.

TABLE 8.1

Typical Information
Technology Costs
per Key Function

Function	Percent of the IS budget
Personnel*	37.75
Hardware	23.25
Software	15.70
Overhead	6.00
Communication	5.55
Networking products	5.25
Outside services	5.10
Other	1.40
	100.00

*As an estimate, this number can be partitioned as 30.75 percent for data processing and 7 percent for data communication.

mately wants to upgrade to ATM at the desktop (a rather unlikely scenario at this time), then the total cost would approach $72 million at today's cost of $1800 per ATM connection; this figure is comparable to 22 percent of the value of the embedded base. If the decision is postponed for a few years, the ATM per-connection cost is expected to decline to $1000 and the total cost would be $40 million, or only 12.5 percent of the value of the embedded base.

The network evolution "wish list" of the organizations we have dealt with recently include the following:[4]

- Adequate bandwidth
- Scalability
- Support of embedded base
- Security
- Technological simplicity

---

[4]As implied in Chap. 7, there is a short-term optimization perspective that focuses on minimizing the number of changes in a network. Although this has short-term appeal, it may not provide the kind of long-term benefits that may be achievable through applying more fundamental changes. For example, if one asked an MIS manager in the late 1980s if he or she thought that a minor tweak to an operating system on the mainframe was preferable to scrapping the mainframe and replacing it with a client/server system, the expected (self-serving) answer would have been obvious. The same situation may be now at play if the same kind of question is asked of the chief networking officer/networking manager of a corporation.

- No changes to applications
- No changes to wiring
- Cost per port from $100 to $500
- In general, overall cost better than with the embedded base or existing alternatives, i.e., 10 to 25 percent cheaper[5]
- No retraining for network management

Communication planners realize that there are many cost factors associated with the running of a network in addition to the basic communication or equipment charges. For example, Table 8.2 identifies some of the so-called hidden factors that have to be accounted for. Although these apply equally well to both ATM and pre-ATM networks, the list makes it clear that the analysis must cover all of these factors as they manifest themselves in the context of each technology under consideration. That is to say, if

$$C_{\mathrm{T}}(tx) = C_{\mathrm{TRAN}}(tx) + C_{\mathrm{EQUIP}}(tx) + C_{\mathrm{ALL\text{-}OTH}}(tx)$$

where
$$
\begin{aligned}
C_{\mathrm{T}}(tx) &= \text{total cost when using technology } x\\
C_{\mathrm{TRAN}}(tx) &= \text{transmission cost when using technology } x\\
C_{\mathrm{EQUIP}}(tx) &= \text{equipment cost when using technology } x\\
C_{\mathrm{ALL\text{-}OTH}}(tx) &= \text{cost of all other factors when using technology } x\\
&\quad (\text{cost of ownership})
\end{aligned}
$$

then a decision based on only a subset of the factors can lead to a suboptimal solution. Table 8.3 provides an example that shows how ignoring certain costs can lead to an incorrect selection.

The specific corporate objective in the design/optimization process is not always the same for all organizations. The optimal design is dependent on the objective. Such an objective could be, for example, to minimize the overall cost of the corporate communication apparatus; or to maximize the quality of service of the network from a performance point of view; or to maximize the profitability of the firm, perhaps by

---

[5]As a matter of course, users imply that mere cost parity with alternatives is not good enough by itself to entice them to upgrade. In reference to ATM, there is evidence that some users are willing to pay incrementally more for ATM, provided they can retain the equipment for a longer period of time because of scalability. Some market research shows that approximately 30 percent of the users would be willing to spend up to 20 percent more on ATM if it provided useful features, including scalability and fault tolerance; however, this implies that the majority, namely, 70 percent of the users, are not willing to pay more.

TABLE 8.2

Hidden Cost Factors Considered by Communications Planners

Business risk

Communication costs:

Initial

Recurring

Cost of capital

Cost of eventual replacement

Cost to back up system, including disaster recovery

Decommission costs

Electric power:

Backup

Planning and administration

Principal

Equipment costs:

Actual equipment

Installation

Cables

Documentation

Fees:

Software/licenses

Licenses

Rights-of-way

Financial management (e.g., accounting)

Lost productivity

Management review

Pilot evaluation

Project feasibility study (staff time)

Project implementation time/cost:

Construction

Cost of service overlap

Delays and overruns

Delivery costs

Project implementation time/cost (*continued*):

Integration

Specialized installation staff

Testing and validating

Project pilot

Rental costs for housing equipment (remote nodes, switches, etc.)

Air conditioning

Floor space

Lighting

Planning and administration

Real estate

Security

Sensors

System/network operation/management:

Ad hoc maintenance

Benefits expenditures

Capacity planning

Cost of documentation

Cost of network/system security (including appropriate tools)

Facilities

Insurance

Maintenance (monthly fees)

Management of staff

Planning and administration

Staff

Test equipment

Training

System/network operation/management:

Turnover

Taxes

Technological obsolescence

TABLE 8.3

Ignoring Some
Factors Can Lead
to an Incorrect
Solution

	C_{TRAN} ($/Month)	C_{EQUIP} ($/Month)	$C_{ALL-OTH}$ ($/Month)	Total ($/Month)
Technology 1	90	150	150	390
Technology 2	130	90	148	368
Technology 3	109	91	170	370
Technology 4	131	151	85	367

Selection when considering only C_{TRAN}—Technology 1. Selection when considering only C_{EQUIP}—Technology 2. Selection when considering C_{TRAN} and C_{EQUIP}—Technology 3. Selection when considering all factors—Technology 4.

maximizing workforce productivity; or to minimize risk of loss and/or maximize safety (for example, for military or public safety departments). Decision makers are well advised to consciously determine what is the objective of a network and to proceed accordingly. When the key objective of a design is to minimize the cost, as often is the case, the methods for comparing alternative strategies call for exact identification of the expenditures associated with each given alternative, as suggested in Tables 8.2 and 8.3.

On a related note, the marketing of technologies by the providers is important. One often hears that technologies such as ISDN, SMDS, and even frame relay were not marketed correctly. Table 8.4 depicts a cost-benefit space that can be applied to various communication services, including ATM. Different vendors and/or vendors' product lines can be mapped to different quadrants of this space to derive a marketing strategy. Some services can be considered "premium" in the sense that they may be expensive, but are very reliable and support many features. Such services may in fact be deemed necessary for some mission-critical components of the overall corporate network. Other networks may be based on "budget" services; if they fail or do not maintain constant end-to-end delay or "block error rate" (we have redefined an old term to encompass frame loss in a frame relay network and cell loss in a cell relay network), no harm will be done. An example could be an E-mail system.

This partition can be used by both the provider and the user. The communication manager should identify both the position taken by the vendor in question and the position that this manager wants to take for the network under consideration. A high-end service may make sense for high-revenue functions, whereas the majority of the users in

TABLE 8.4

Service Priorities:
Positioning
Telecommunication
Services

	Fewer Features	Baseline Features	More Features
Premium (more expensive)	Undesirable/ unfeasible/ unsustainable	Undesirable/ unfeasible/ unsustainable	High-end applications/ power users
Standard (baseline price)	Somewhat undesirable (some ATM switch vendors fit this profile)	Normal situation/ sell	Desirable (vendor has no problem selling)
Budget (less expensive)	Possible strategy (e.g., Kmart)	Desirable	Great/easy sell

the organizations may in fact be more cost-effectively supported by other less advanced technologies. For example, the campus-level ATM service LANE (Chap. 7) allows ATM-based PCs and workstations connected to an ATM hub that is connected to an ATM-to-Ethernet converting bridge to be internetworked with an Ethernet-based population. This enables the critical people in the organization—for example, the "traders" of financial instruments—to have ATM in support of features such as multiwindow video, imaging, and bulk file transfer without having to incur the cost of upgrading the entire organization.

From the technology provider's point of view, this partition of the space is useful in identifying the marketing, sales, and application analysis strategy. For example, a high-end service should not be marketed to an entry-level individual; however, such an individual may well be able to make a decision concerning the purchase of a $79 modem or a low-end service such as an extra dial-up line. A high-end service probably will not be accepted for a low-end, low-bandwidth, low-revenue, tactical application; it should instead be positioned for the high-visibility, new-technology, strategic applications.

In summary, what determines which communication services are chosen by user organizations, particularly with regard to ATM? Many researchers have found that ATM end-user demand is a price issue and that in the immediate future, ATM deployment will be greatly influenced by its economics. The following are perhaps the key observations:

1. The major corporate IS investment is in application software. The network interface cards (NICs) are not the only/primary consideration. A new service must support this software base.

2. In general, savings over the present mode of operation must be 15 to 20 percent or better.

3. Designers consider the entire corporate topology, not just a few discrete links, in selecting the best solution. Network mangers typically prefer to deploy a smaller set of technologies, namely two or three distinct technologies at most in each category (e.g., transmission)—not five or ten technologies.

4. Transmission cost is important, but there are other hidden costs to be taken into account. Examples include security, reliability, maintainability, consistency with standards, interoperability, etc.

5. In the LAN there are many alternatives available for desktop connectivity. Because of the size of the population and the plethora of available technologies, it will be more difficult to justify ATM at this time. In the WAN, ATM is the only available switched broadband QoS-guaranteed service; if broadband is a requirement, it should be fairly easy to justify ATM. The same applies to the campus backbone.

8.3   ATM Deployment Options

The economics of ATM have to be considered separately at the LAN and WAN levels, because they are different. Organizations have contemplated a number of ways in which they can deploy ATM in their enterprise networks. Some of the key options are

1. *ATM to the desktop for all users.*   This would require ATM-configured hubs and, very likely, premises ATM switches; WAN ATM services may also be used.

2. *ATM to the desktop for some users.*   LAN emulation service capabilities are then required.

3. *ATM to the hub; ATM in backbone and Ethernet (possibly switched Ethernet) to the users.*   Hubs may be directly interconnected or may employ ATM premises switches.

4. *ATM in the campus backbone only; Ethernet hubs terminate on a single premises ATM switch.*   The switch provides interworking function between Ethernet and ATM. Interworking options include simple Ethernet encapsulation (e.g., RFC 1483), classical IP over ATM, LAN emulation service, and MPOA.

5. *ATM in the campus backbone only; Ethernet hubs terminate on a group of premises ATM switches that are connected over links with private network-node interfaces.* Switches provide interworking function between Ethernet and ATM. Again, interworking options include simple Ethernet encapsulation (e.g., RFC 1483), classical IP over ATM, LAN emulation service, and MPOA.

6. *ATM in the campus backbone only; Ethernet hubs terminate on routers that use ATM between them (over a campus network).* Again, interworking options include simple Ethernet encapsulation (e.g., RFC 1483), classical IP over ATM, LAN emulation service, and MPOA.

7. *ATM in the WAN backbone only; Ethernet hubs terminate on routers that use a public ATM service between them (over a WAN network).* Most likely, interworking options include simple Ethernet encapsulation (e.g., RFC 1483). Other interworking approaches, such as classical IP over ATM, LAN emulation service, and MPOA, have to be implemented privately by the user because the carrier probably will not support these functions in the network. They can then tunnel through the public ATM network.

Each of these options involves distinct economic and design considerations. In addition to the topologies just described, one needs to take into consideration which ATM service is of interest, e.g., permanent virtual connections, switched virtual connections, constant bit rate, variable bit rate, available bit rate, and unspecified bit rate. Furthermore, one needs to consider whether the network carries only data or carries voice, video, and multimedia. In general, voice over ATM is not cost-effective compared to available alternatives; video over ATM may be cost-effective in some situations; data over ATM can be cost-effective in many situations.

8.4   ATM Equipment and Costs

This section addresses ATM equipment costs. From a functionality point of view, ATM equipment can be grouped as follows:

- ATM access equipment [e.g., network interface cards (NICs), router boards, server boards, data service units]

- ATM switching equipment (e.g., ATM floor hubs, private ATM switches, carrier edge switches, carrier core switches)

- ATM transmission systems equipment (e.g., premises wiring systems, fiber-based local loop equipment including SONET rings, interoffice transmission equipment)

- ATM interworking equipment (e.g., LAN emulation servers, MPOA route servers, carrier service nodes)
- ATM testing and network management systems and equipment
- ATM video and multimedia equipment (corporate and public video servers, set-top boxes)

As noted earlier, ATM equipment is now reaching the market at an expedited pace. Table 8.5 gives a partial list of premises ATM equipment vendors, whereas Table 8.6 lists vendors that provide public network equipment. ATM technology is now experiencing improved cost/performance figures, not only because of increased performance but also because of decreased equipment costs. For example, ATM switch ports on carrier/workgroup switches have gone from $3000 per port in early 1993 down to $600 in early 1997 in some instances (e.g., 25-Mbps ATM); workstation adapters went from $4000 in early 1993 down to as little as $222 in early 1997 (e.g., 25-Mbps ATM).

Note: All prices mentioned in this chapter are budgetary list price. Volume discounts in the 20 to 40 percent range are achievable by large users.

Note: Device prices depend on the bus used, the medium (multimode, single-mode, coax, unshielded twisted pair), the speed (DS1, DS3, OC-3, OC-12), the scope (LAN/WAN), and even the power budget range. Hence, single-number generalizations for the price of these elements are not possible, and are discouraged to the extent possible.

8.4.1 PC/Workstation Adapters

As noted, these devices have experienced improvement in their cost-effectiveness in the recent past. For organizations contemplating ATM all the way to the desktop, adapter costs can be a significant expense item because of the multiplicative implication: The cost is directly proportional to the number of upgraded PCs and workstations. Current pricing ranges from $222 to $1300 (higher in special cases) (see Table 8.7). Note that the price depends on the network-side speed (e.g., 25 Mbps, 155 Mbps, etc.), the medium (UTP, single-mode, multimode), the memory on the board, and the kind of bus-side interface (e.g., PCI, SBus, EISA, VME, GIO, PMC, etc.). Generally, the cost of the hub termination is comparable to the PC adapter cost. At this time, most organizations have made the transition at the desktop to the use of (1) switched Ethernet, (2) 100-Mbps Ethernet, or (3) switched 100-Mbps Ethernet. ATM to the end device is reserved for either high-end servers or select digital production (e.g., imaging, video) applications.

TABLE 8.5

Vendors Providing User-Level ATM Products (Partial List)

3Com	Lucent Technologies
Adaptive Corporation	NEC America
ADC Fibermux	NetEdge (Larscom)
ADC Kentrox	Network Equipment Technologies
ADC Telecoms	Network Peripherals
Agile Networks	Network Systems
Alcatel	Newbridge Networks
Bay Networks	NiceCom
Cabletron	Nortel
Cascade/Ascend	Odetics
Cisco/Stratacom	Olicom
Connectware/AMP	Optical Data Systems
Digital Equipment	Retix (Sonoma)
Digital Link Corp.	Sun Microsystems
FastComm	SuperNet
Fibercom	Synemetics
Fore Systems	Telco Systems
Fujitsu	Telematics (ECI)
General DataComm	Tellabs
GTE Government Systems	TRW
Hughes LAN Systems	Ungermann-Bass
Hughes Networks	Xylan
IBM	Xyplex
Interphase Corp.	

8.4.2 Access Multiplexers

These devices support multiplexing of data, voice, and video, and occasionally service interworking (for example, they convert packet/Ethernet traffic to ATM). They typically have a backplane speed of 0.2 to 2 Gbps and support a variety of access speeds [56/64 kbps, FT1/E1, 1.5 Mbps, 45 Mbps (ATM, SMDS, HSSI), EC-3]. They range in price from $10,000 to

TABLE 8.6

Partial List of Companies Supplying Carrier-Level ATM Products

ADC Telecommunications	General DataComm
Alcatel Network Systems	Lucent Technologies
Cascade (Ascend)	NetEdge (Larscom)
Cisco/Stratacom	Nortel
DSC Communications	Siemens
Fore	Stromberg-Carlson
Fujitsu Network Switching	Telematics (ECI)

TABLE 8.7

Typical Adapter Pricing (e.g., Interphase Corp.), 1997

PCI 155-Mbps SONET OC-3, multimode, 128K RAM	$850
PCI 155-Mbps SONET OC-3, multimode, 512K RAM	$1050
PCI 155-Mbps SONET OC-3, multimode, 1M RAM	$1225
PCI 155-Mbps SONET OC-3, UTP, 128K RAM	$650
PCI 155-Mbps SONET OC-3, UTP, 1M RAM	$1025
PCI 25-Mbps, UTP, 128K RAM	$222
SBus 100-Mbps TAXI, multimode, 1M RAM	$975
SBus 155-Mbps SONET OC-3, multimode, 1M RAM	$995
SBus 155-Mbps SONET OC-3, UTP, 1M RAM	$795
EISA 155-Mbps SONET OC-3, multimode, 512K RAM	$1095

$45,000. The range of equipment in this category is still relatively limited. Hopefully, more vendors will enter the market in the future.

8.4.3 Layer 2 Switches

During the 1990s there has been significant penetration of Layer 2 switches supporting switched 10/100-Mbps LANs. These switches, in turn, need to be connected over a campus/WAN backbone. Table 8.8 provides a sense of the cost of this technology at this time.

8.4.4 ATM Interface Cards for Routers

Routers are playing a key role in organizations' migration toward ATM. Backbone use of ATM, whether at the campus or WAN level, has

TABLE 8.8

Typical ATM-ready
Layer 2 Switches
(e.g., Cisco), 1997

Catalyst 5000	
Functional chassis	$11,900
24-port 10BaseT Ethernet switching module	$9,900
10/100BaseTX Fast Ethernet switching module	$9,900
FDDI switching module	$9,900
100BaseFX switching module (SM)	$19,900
ATM LANE (UTP)	$8,900
ATM LANE (SM)	$14,900
ATM LANE (MM)	$9,900

TABLE 8.9

WAN Hardware for
Routers (e.g.,
Cisco), 1997

Cisco 7000/7500 ATM Interface Processor	
100 TAXI MMF	$20,000
155-Mbps SONET MMF	$22,000
155-Mbps SONET SMF	$25,000
E3 coax	$18,000
DS3 coax	$18,000
Cisco 4000/4500 ATM Interface Processor	
155-Mbps SONET MMF	$9,000
155-Mbps SONET SMF	$11,000
E3 coax	$12,000
DS3 coax	$12,000

emerged as a mainstream application of ATM. Table 8.9 gives prices of typical WAN router cards (these prices are based on Cisco's pricing, as this company has more than 75 percent of the market). In addition to using ATM as a WAN discipline, routers are also using ATM internally as a way to switch traffic more efficiently. The IP Switching approach is based on this concept. MPOA is also driving a redesign of routers to make better use of ATM as a single-hop connection strategy, rather than utilizing the current method of multiple end-to-end hops over dedicated private lines. MPOA servers supply the address of the direct route over an ATM channel for the Layer 3 IP packet to flow so that the connection is set up end to end. Much more development is expected here in the next couple of years.

8.4.5   Premises/Carrier Switches

Two categories of products are now reaching the corporate market: (1) midsize/large-scale enterprise switches (used in campus and wide area environments), and (2) small workgroup switches for departmental LAN applications. The typical per-port cost for these ATM switches, which can be used for campus or small-carrier applications, is illustrated in Table 8.10. Although this table shows list prices for Fore, they are similar to the prices of competitive switches by vendors such as Cisco, Ascend/Cascade, Newbridge, GDC, and Telematics, to list just a few.

8.4.6   Testing Equipment

Testing is very important, as can be inferred from the popularity of LAN protocol analyzers and testers. These testers continue to be expensive and range in price from $25,000 to $100,000 (lab-level testers for more elaborate network management systems are even more expensive). Fortunately, more vendors are entering this market, and new, less expensive products should become available in the next 24 months.

8.4.7   Network Services

At this time ATM is penetrating the WAN as a technology to support higher-speed connectivity than is possible with frame relay services or dedicated point-to-point DS3 lines. The desire to increase connectivity to all of the corporate sites (thereby increasing the number of points on the network) and to increase the throughput of the Internet connecting Layer 2 switches supporting either switched Ethernet or 100-Mbps switched Ethernet is driving the introduction of ATM services in the WAN. The price of these services comprises, generally, the following components: scope (metro versus nationwide), port charge, throughput charge, and PVC charge (number of PVCs); also, there are installation charges. Typically, these services are priced aggressively compared with the point-to-point dedicated-line equivalent. Table 8.11 provides a gamut of current information on available carriers, geographic coverage, service types, and typical pricing strategies.

TABLE 8.10

Typical ATM Switch Pricing (e.g., Fore), 1997

Fore ASX-200 WG (2.5-Gbps workgroup switch, 1 power supply, 4 network module slots) (workgroup application)	
Chassis with 12 ports of OC-3 MMF	$19,900
Chassis with 16 ports of OC-3 MMF	$21,900
Chassis with 12 ports of 155-Mbps UTP	$18,900
Chassis with 16 ports of 155-Mbps UTP	$20,900
Chassis with 24 ports of TAXI 100	$23,900
Chassis with 18 ports of 25-Mbps UTP	$9,900
Fore ASX-200BX chassis only (2.5-Gbps switch, 2 power supplies, 4 network module slots) (carrier or workgroup application)	$21,900
Fore ASX-1000	
2.5-Gbps chassis	$33,900
5-Gbps chassis	$49,900
7.5-Gbps chassis	$57,900
10-Gbps chassis	$69,900
6-port WAN T1 ATM UNI	$14,900
6-port WAN E1 ATM UNI	$18,900
4-port WAN J2 ATM UNI	$24,900
4-port WAN E3 ATM UNI	$17,900
4-port WAN DS3 ATM UNI	$13,900
4-port WAN OC-3 ATM UNI	$24,900
1-port WAN OC-12 ATM UNI	$13,900
6-port LAN 25-Mbps UTP	$1,700
6-port LAN 100-Mbps TAXI	$4,900
4-port LAN 155-Mbps MMF	$3,400
4-port LAN 155-Mbps UTP	$2,900
1-port LAN 155-Mbps SMF short-reach, 3-port 155-Mbps MMF	$5,900
1-port LAN 155-Mbps SMF long-reach, 3-port 155-Mbps MMF	$12,900
1-port LAN 622-Mbps MMF	$7,900

TABLE 8.11 Typical ATM WAN Services Pricing (Early 1997)

Carrier	Service Date	Locations	UNIs	Service*	Pricing Strategy	Type	NRC	MRC
Ameritech	1994	6 major metros in northeast	DS1–OC-12	CBR, VBR	Customer-specific	Flat, distance-sensitive	$2000	DS1: $1000; DS3: $4000; OC-3: $4700
ACSI	1996	34 major metros in east, southeast, southwest	DS1–OC-3	CBR, VBR, UBR	Customer-specific	Flat, metro not distance-sensitive/ WAN distance-sensitive	NA	Local access, port, and service class fee
Bell Atlantic	1996	Philadelphia and Pittsburgh	DS1–OC-3	CBR, VBR	Customer-specific and tariffed	Flat, distance-insensitive	$1500 (DS3); $2500 (OC-3)	DS3: $3500–4400; OC-3: $7500–9250
BellSouth	1995	North Carolina and Knoxville	DS1–OC-3	CBR, VBR, UBR	Customer-specific and tariffed	Flat, distance-sensitive	$850	DS1: $1550; DS3: $2160; OC-3: $3650
GTE	1994	About a dozen major metros nationwide	DS1–OC-3	CBR, VBR, UBR	Customer-specific and tariffed	Flat, distance-insensitive	$3000	$40/Mbps plus $10/Mbps for VBR PCR
Pacific Bell	1993	5 major metros in California	DS1–OC-3	CBR, VBR, UBR	Customer-specific and tariffed	Flat, distance-sensitive outside metro	$400 (DSI); $1500 (DS3); $3000 (OC-3)	DS1 (CBR): $792; DS1 (VBR): $672; DS3 (CBR): $4399; DS3 (VBR): $3799; OC-3 (CBR): $9176; OC-3 (VBR) $6956
Southwestern Bell	1996	12 cities in 5-state region	OC-3	CBR, VBR	Customer-specific	Flat, distance-sensitive outside metro	$600 DS3 or OC-3	DS3: $2800; OC-3: $3300
Sprint	1993	Florida, North Carolina, South Carolina, Virginia, Missouri	OC-3–OC-12	CBR, VBR	Customer-specific and tariffed	Usage-sensitive, distance-insensitive	NA	NA
TCG	1995	20 major cities (New York, Boston, Chicago, Houston, Dallas, Pittsburgh, Phoenix, San Diego, San Francisco, Los Angeles, Seattle)	DS1–OC-12	CBR, VBR, UBR	Customer-specific	Flat, distance-insensitive	$3000 DS3 or OC-3	DS1: $1590–1749; DS3: $4935–5483; OC-3: $8290–9223
US West	1995	Minnesota, Oregon, Washington	DS3–OC-3	CBR, VBR	Tariffed	Flat, distance-insensitive	$1200	Port: $440; access link DS3: $648–710; access link OC-3: $967–1069

*CBR=constant bit rate; UBR=unspecified bit rate; VBR=variable bit rate.

8.5   Future Cost Trends for ATM Equipment

This section documents some anticipated pricing trends based on primary and secondary market research as well as industry data.

PC/workstation ATM NICs.   Our analysis shows the following NIC cost estimate (averaged over the 25-, 52-, 100-, and 155-Mbps values): $850 in 1997; $800 in 1998; $700 in 1999; $600–400 after that.

ATM hub cards to support the workstation/PC.   These cards are slightly more expensive, as they may incorporate more sophisticated network management features and/or may need to operate at a higher speed on the bus side. Our analysis shows the following cost estimate (averaged over the 25-, 52-, 100-, and 155-Mbps values): $950 in 1997; $900 in 1998; $700 in 1999; $600–500 after that.

Switch ports on enterprise switches.   Our analysis shows that WAN DS3 ports with full ATM functionality (including traffic control, switched virtual connections, network management, and perhaps support for multipoint connectivity) will probably follow the following cost trajectory: $4000 in 1997; $3800 in 1998; $3600 in 1999; $3400 in 2000. (These figures refer to the port on the line card itself, and not to the prorated portion of the entire switch; also note that, in general, each line card supports 4, 6, or 8 ports, making the card comparably more expensive.) WAN OC-3 port prices are as follows: $6000 in 1997; $5500 in 1998; $5000 in 1999; $4500 in 2000.

Router-to-network ports.   Our analysis shows that these ports are expected to cost $10,000–20,000 in 1997 and $8000–16,000 in 2000.

8.6   Competitive Technologies

As implied by the discussion above, there are now several technologies competing with ATM, particularly at the premises/campus level (see Table 8.12). Ultimately, the success of one technology or another will depend on features and capabilities and on the cost; cost is particularly important for those items that have high-replication implications (e.g., NICs).

While many alternatives are available at the building level, fewer alternatives are available at the campus level, and even fewer at the WAN level.

TABLE 8.12

Competing Technologies at the Campus Level

Switched Ethernet	Switched Ethernet supports microsegmentation of LANs into segments supporting a few users or just one user. Additionally, it enables the interconnection of these segments into a network supporting any-to-any communication. The technology provides dedicated 10-Mbps bandwidth to each user. This practical technology extends the useful life of Ethernet, particularly for nonvideo applications (some forms of video can also be supported where needed). Switching hubs support several hundred Mbps and 10 to 25 Ethernets. Prices are approximately $800 per connection.
100-Mbps Ethernet	Provides 100-Mbps on shared medium using Ethernet access discipline. The technology is simple and requires no changes to applications or to the premises wiring. The cost per port is $400 to $700. Limitations include the use of shared bandwidth, which is not ideal for multimedia, and the fact that it is not scalable. Also, there are distance limitations.
1000-Mbps Ethernet (Gigabit Ethernet)	An evolution of Ethernet technology that provides up to 1 Gbps bandwidth for desktop applications. Physical systems supported include shielded twisted pair (1000BaseCX, 25 m), unshielded twisted pair (1000BaseT, 100 m), short-wavelength optics (1000BaseSX; 550 m), and long-wavelength optics (1000BaseLX; 3000 m). With the exception of the 1000BaseT, the 8B/10B encoding used in Fibre Channel is utilized. The IEEE 802.3z standard was being targeted for the first quarter of 1998.
FDDI	Provides 100 Mbps on shared medium using token-passing access discipline. The technology is expected to be a major contender for enterprise high-speed campus/building backbone networks until about 1997 (about half of the new campus networks being deployed still employ FDDI based on product maturity, stability, and reliability); it is not widely deployed at the desktop. The limitation of FDDI is that it is, mostly, a data-only/campus-only solution that does not scale upward in bandwidth. Copper-based FDDI (CDDI) NICs range in price from $650 to $950.
Other	A number of other technologies, among them the Fibre Channel standard, also compete with ATM at the local area network level.

This is the reason why ATM is expected to see good penetration at the WAN and campus levels in the next couple of years.

8.7 Conclusion

ATM has become a "prime time alternative" in almost all segments of an enterprise network, with the exception of the desktop component. There is a plethora of equipment available from mainstream vendors such as Cisco, Bay Networks, Fore, Ascend/Cascade, Newbridge, and many others. ATM capabilities have been added to uplinks for Layer 2 switches (e.g., LAN switches), to WAN links for routers, and to workgroup switches. ATM is available as a WAN service from many carriers, either as native cell relay service or as a way to transparently interconnect legacy enterprise LANs at native speeds. ATM is being deployed in the Internet, and there is even discussion of using ATM for voice applications (Chap. 11). Significant penetration of this technology is anticipated in the next few years.

8.8   Reference

[dmdp] D. Minoli, *ATM Economics*, DataPro Report, McGraw-Hill, May 1997.

ATM
Network Design
Considerations

As discussed in previous chapters, asynchronous transfer mode technology and services for both local area/campus and wide area network applications are now being brought to the market, and also in support of IP. A plethora of equipment in support of enterprisewide (data) applications is becoming available from a relatively large number of vendors; in addition, over a dozen carriers have introduced or announced cell relay services (Chap. 8).

ATM has reached technological stability, at least for traditional enterprise applications. The significant questions now facing users relate to the cost-effective incorporation of ATM into the embedded base of existing corporate networks and subnetworks. Prospective users now face issues pertaining to network design considerations and tools. This important phase of the maturation of ATM will determine if, how, and how rapidly the technology will experience widespread deployment in mainstream, mission-critical environments. These design considerations will facilitate the analysis that network planners must undertake in order to satisfy financial "due diligence" requirements. These requirements are intended to ensure that all pertinent alternatives are considered, and that the most efficient and cost-effective communication technology is chosen. Business process reengineering (BPR) in major corporations will continue for at least several more years, ensuring that rigorous cost analysis of new technology will be required before corporations purchase any new systems or services. However, technology in general and broadband in particular are expected to figure prominently in the U.S. corporate landscape, and in corporations' productivity plans and competitive machinery, as we approach the new century.

Design considerations address the following issues: topological design (when/where to use ATM technology?), performance (what are the end-to-end delay, jitter, and loss?), and capacity (when do new resources—switches and communication facilities—have to be added to meet the expected grade of service?). Table 9.1 depicts some of the performance metrics of interest. They tend to be driven by network engineering and capacity planning.

As they pursue savings, organizations have become increasingly unwilling to automatically replace existing communications equipment and services with the latest generation of (more expensive) technology without careful analysis of the technology's benefit. Network design considerations and capabilities become increasingly critical in new technology purchase decisions.

This chapter addresses some of these important design issues [dmdp]. The first section examines pertinent network design considerations. The

TABLE 9.1

Network Performance Metrics of Interest in ATM (and Other Networks)

End-to-end delay. This is the total delay of information across the network or between particular endpoints; it may include transmission delays, propagation delays, queueing delays, processing delays, and switching delays.

Throughput. This is the amount of information per second transported by the network or network component; it may be measured across the entire network, from endpoint to endpoint, on a link, or across a switch.

Jitter. This is the variability of end-to-end delay; it results from changes in constituent delays from cell to cell.

Cell loss ratio (CLR). This represents the number of cells dropped by a network or component as compared to the total number of cells submitted; it results from congestion and transmission (bit) errors. This factor can be measured across the network or across a network component.

Connection block ratio. This is the number of connections not accepted as compared to the number requested.

Link utilization. This is the proportion of link capacity actually employed to transmit cells.

Switch utilization. This represents the proportion of switch capacity actually used to process cells.

section that follows looks at product capabilities and discusses features of some of the available tools. Specific corporate decisions regarding ATM deployment can be made only in the context of an organization's unique communications environment. Nonetheless, the generic observations presented in this chapter serve as an initial input to the necessary analysis.

9.1   Design Considerations

Enterprise communications network design typically covers the LAN segment, the campus segment, and the WAN segment. When deploying ATM in their networks, some companies may choose to restrict ATM to the LAN and campus segments, retaining traditional equipment and services in the WAN segment. Other companies will plan to use ATM only in the WAN. A smaller group will consider using ATM end to end. These three approaches lead to three fairly distinct design challenges, since different design and cost issues come into play in different segments. Also, there is an additional distinction based on whether the ATM WAN is private or uses public services.

9.2 Local Area ATM Network Design

In LAN and campus environments, objectives center on increasing performance cost-effectively. When considering introducing ATM into their LANs and campus networks, designers must decide whether to deploy ATM to all users' desktops; bring ATM only to certain users, supporting the remainder via LAN emulation services; or use ATM only in the backbone, with no direct user ATM connections. Planners must also settle on the type of cabling to employ, whether unshielded twisted pair (UTP) category 5, UTP category 3, or multimode fiber. Chapter 8 listed some of the available alternatives. In designing the backbone portion of the network, designers must decide whether to use directly interconnected ATM-configured hubs, directly interconnected ATM-configured routers, or premises ATM switches. Each combination of components and strategies has specific performance improvement and cost implications; however, once deployed, the networks incur no ongoing communication costs.

In a typical downtown financial services BAN (building area network), a number of risers interconnect traditional Ethernet hubs to two backbone ATM switches. This design supports local and centralized file and application servers. Routers are used to provide a precautionary firewall, helping to keep traffic from existing, non-ATM servers, such as Novell NetWare servers, from flooding the network. Two noncolocated switches can be used for reliability and load sharing, and hubs are dual-homed to the switches, assuring connectivity in case of the failure of one switch. The backbone forms a loop across nonintersecting physical conduit to eliminate the possibility of a single point of failure. Once installed, this system will provide much greater bandwidth than traditional networks, provide high reliability, and support bandwidth-intensive applications.

9.2.1 LAN Design Issues

The three design issues listed at the beginning of this chapter have applicability at the LAN level, although these issues are more acute in WANs.

Topological design (when/where to use ATM technology?).   At this juncture, the most cost-effective use of ATM in the LAN environment is for campus/riser backbone support (e.g., replacing FDDI-based backbones, as exemplified in the system just discussed). Typically, local routers or Layer 2 switches (i.e., 10/100-Mbps Ethernet switches) can be interconnected via ATM-ready uplinks, usually through the use of local multimode fiber. For small networks (e.g, a few backbone routers), the aggregation devices (e.g., routers) can be connected directly over point-to-point ATM links. When there are more than a few devices (say, more than a dozen), then the use of a campus ATM switch, to which these aggregation devices are connected, is strongly encouraged. Beyond basic topological considerations, there are considerations regarding protocol architectures. Many enterprise networks today are based on IP and on routers. Therefore, key questions are, "How to carry IP over an ATM (backbone) network?" and "How to utilize legacy routers in the new environment?" The first question has four answers, as follows (also see Chap. 7):

1. Utilize devices that implement RFC 1577, classical IP over ATM, which is a mechanism to set up tunnels over the ATM network to carry IP packets.

2. Utilize devices that implement the LAN emulation (LANE) specification of the ATM Forum (along with RFC 1483), which enables interworking between ATM devices and Ethernet/token ring LANs by applying translational bridging principles (translating between the connectionless LAN environment and the connection-oriented ATM environment).

3. Utilize devices that implement the multiprotocol over ATM (MPOA) specification of the ATM Forum, which enables routers to cut through to the destination using a single hop over an ATM network and supports both bridging and routing methods.

4. Utilize any of the Layer 3 switching (L3S) technologies, such as Tag Switching, IP Switching, or Multiprotocol Label Switching (MPLS), to support more effective movement of IP protocol data units (PDUs) across the network.

The second question has answers that parallel the answers to the previous question:

1. When using the RFC 1577 method, existing routers have to be upgraded at the software level to support the RFC and at the hardware level to support ATM.

2. When using LANE, existing internetworking devices have to be upgraded at the software level to support LANE, including the LAN emulation server, the broadcast and unknown server, and the

LAN emulation and configuration server, and at the hardware level to support ATM.

3. When using MPOA, existing routers have to be upgraded at the software level to support the specification and at the hardware level to support ATM.

4. When using L3S, appropriate software/hardware upgrades are needed to routers and switches.

Performance (*what are the end-to-end delay, jitter, and loss?*).   This question relates to (1) the service class that may be utilized by the router's point-to-point ATM link or by the campus switch used to interconnect the routers, (2) the kind of congestion control utilized by the ATM hardware, and (3) the engineering/overbooking policy on the ATM switch. Generally these considerations apply more to WAN environments, as long as the campus switch is not overutilized and/or the WAN link (if any) is not oversubscribed.

Capacity (*when do new resources—switches and communication facilities—have to be added to meet the expected grade of service?*).   This question relates to (1) the switch throughput (generally not a problem, since many switches are nonblocking—e.g., if it supports 16 OC-3 ports, then it has a 2.5-Gbps backplane); (2) the multiplexing strategy used, e.g., how many virtual channels (VC) of a given sustainable cell rate (SCR) and peak cell rate (PCR) originating over a number of input ports are mapped into an outgoing port of given maximum port speed; and (3) the switch-to-switch link capacity, particularly as related to the multiplexing strategy used to map many VCs originating over a number of input ports to the outgoing switch-to-switch link; and (4) the engineering/overbooking policy on the ATM switch.

9.2.2   Traffic Management

Traffic management (TM) encompasses some of the factors that have been alluded to in this LAN/campus design discussion, but that play a more pronounced role in WAN design. Traffic management in ATM networks consists of those functions that ensure that each connection receives the Quality of Service it needs and that the flow of information is monitored and controlled within the ATM network. ATM networks use three techniques to manage traffic: traffic shaping, traffic policing, and congestion control.[1]

---

[1]This section is based on Ref. [fo].

Traffic Shaping   Traffic shaping is a management function performed at the user-network interface of the ATM network. It ensures that traffic matches the contract negotiated between the user and the network during connection establishment. Traffic is shaped according to the Generic Cell Rate Algorithm (GCRA) specified by the ATM Forum UNI standard Version 3.0. Devices implementing traffic shaping are typically those connected to an ATM network and include ATM network adapters in PCs or workstations, hubs, bridges, routers, and data service units (DSUs).

Traffic Policing   Traffic policing is a management function performed by the ATM network (i.e., ATM switches) that ensures that traffic on each connection remains within the parameters negotiated at connection establishment. To police traffic, ATM switches use a buffering technique called a leaky bucket. It is a system in which traffic flows (leaks) out of a buffer (bucket) at a constant rate (the negotiated rate), regardless of how fast it flows into the buffer. The need for policing occurs when traffic flow exceeds the negotiated rate and the buffer overflows. The ATM switches must then take action to control (police) it. In the header of each ATM cell is a bit called the cell loss priority (CLP) bit. The ATM switches use this bit to identify cells as either conforming (to the contract) or nonconforming. If cells are nonconforming (for example, if there are more cells than the contract allows), the ATM switch sets the CLP bit to 1. This cell may now be transferred through the network only if the current network capacity is sufficient. If it is not, the cell is discarded and must be retransmitted by the sending device. Constant bit rate (CBR) traffic requires a single buffer (leaky bucket) to police the traffic because CBR traffic uses only a sustained (average) rate parameter in its network contract. Variable bit rate (VBR) traffic uses two buffers (dual leaky buckets) to monitor both the sustained rate over a discrete time period and the maximum (peak) bandwidth used during the connection. If either value exceeds the contract parameters, the ATM switch polices the VBR traffic by manipulating the CLP bit. Traffic policing can be used to control overbooking of outgoing trunks, thereby being important to design considerations.

Congestion Control   In a well-designed ATM LAN, CBR and VBR traffic experience the low-latency service they negotiated at setup, while available bit rate (ABR) traffic might experience congestion, depending upon the current loading of the network. Since the applications that use ABR connections are less sensitive to delay, all applications run as planned. Congestion control is needed because the ABR traffic is likely to experience congestion at some point in time. If the congestion is controlled, the ABR service still provides value. CBR and VBR are designed to require no complex congestion management schemes. Several schemes have been proposed, and the two leading schemes involve controlling traffic flow on either a link-by-link basis or an end-to-end basis. The ATM Forum worked on a solution that integrates the best features of each.

End-to-End   End-to-end flow control is readily available from most vendors and at a relatively low price. But it has two major drawbacks: Recent studies show that cells can be lost during congestion, and it requires a lot of buffer space. End-to-end schemes control the transmission rate at the edge of the network, where the LAN meets the ATM device. If the ATM-to-LAN connection is Ethernet to the WAN, at which point an access device converts the traffic to ATM, then this is a low-cost method of connecting to the WAN. Because little of the LAN beyond the backbone uses ATM technology at this time, precise control may not be required, and the few extra buffers will not substantially increase the cost. As ATM equipment prices continue to drop, however, ATM technology will be used in more of the LAN. Paying for the extra buffer space that will be needed will be expensive, making precise control over more complex networks a very important issue. Thus, the compromise is to include the option of providing more precise link-by-link control.

Link-by-Link   Link-by-link flow control supports more users but uses less buffer space, all without losing cells. Link-by-link has two major drawbacks at first: It will be more expensive, and equipment that implements it is not yet available. Link-by-link schemes, besides providing control on a per-link basis, also control each VC separately. This allows the network of ATM switches to control a particularly active device, while other devices continue receiving a fair share of the available network through a rate-based method or a credit-based method. The rate-based method controls the flow of traffic by communicating to the sending device the rate at which it can transmit (the allowed rate). The credit-based methods control traffic flow by communicating to the sending device the remaining buffer space (credits) the downstream device (receiver) has to receive traffic, on a per-VC and per-link basis. ABR will use the rate-based method.

Integrated   The integrated proposal provides an end-to-end, rate-based scheme as the default method, with the link-by-link scheme as an option. Because most equipment can provide this level of control immediately, users will have a standards-based ABR congestion control scheme very quickly. If the network requires more precise congestion management, the optional scheme can be used to increase the control of the ABR resources. This would still be a standards-based solution, with end-to-end and link-by-link equipment coexisting in the same network. When a connection is made from an end-to-end device, the link-by-link device would simply perform the end-to-end control scheme when talking to that device. This would preserve the existing investment in equipment, while providing for future growth.

9.3   Wide Area ATM Network Design

Wide area network design objectives differ considerably from LAN design objectives. When designing and implementing a wide area ATM network, designers seek to (1) minimize the cost of the network over time, given that there will be recurring and/or usage-sensitive charges, (2) satisfy reliability and route diversity requirements, and (3) satisfy performance requirements. Design issues include the architecture selection—whether to deploy a private, public, or hybrid network—equipment selection, carrier selection, and design tool selection.

9.3.1   Public Network Design

When selecting a *public* cell relay service, the user needs to make a relatively small number of design decisions. One selection relates to the access portion of the network. Access to a nationwide ATM/cell relay service can be accomplished in three ways:

1. Use a cell relay service provided by a local exchange carrier (LEC) or regional Bell operating company (RBOC) that interworks with an interexchange carrier (IXC) cell relay service.

2. Use a dedicated LEC/RBOC-provided facility to an IXC's point of presence (POP), where the IXC's cell relay service can be obtained. The dedicated facility can be a T1 or T3 line; a synchronous optical network (SONET) electrical carrier-1 (EC-1), EC-3, or EC-12 service; a "dark fiber" link to the POP; or a set of T1/T3 lines that are inverse-multiplexed to obtain a higher-speed line.

3. Use an alternate access provider (AAP) service to connect to a traditional IXC or for end-to-end service where the AAP provides

long-distance facilities. The two largest providers at this time are Teleport Communications Group and MFS.

9.3.2 ATM Service Classes Used in Design

Carrier selection will depend to some degree on the class of service and the Quality of Service (QoS) required for the application or applications that the organization has in mind. Not all carriers offer all service classes specified for ATM. ATM service classifications and traffic classes were discussed in Chap. 2.

The original ATM Forum specifications (UNI 3.0 and UNI 3.1) supported only CBR and VBR. The major improvement of the ATM Forum TM 4.0 specification over previous UNI specifications was the addition of the ABR text and the clarification of the service categories [specifically, the distinction between real-time (rt) VBR and non-real-time (nrt) VBR, and also unspecified bit rate (UBR)]. Prior to the TM 4.0 work, the only text addressing traffic management was contained in the original ATM Forum UNI 3.0 specification. The UNI 3.0 document was primarily concerned with congestion avoidance mechanisms, whereas the TM 4.0 document contains language that addresses how to provide differentiated QoS more efficiently [as].

It is worth noting that prior to the TM 4.0 work, traffic management at call setup time was governed by the users' combination of parameters: PCR, cell delay variation tolerance (CDVT), SCR, maximum burst size (MBS), CLR, cell transfer delay (CTD), and cell delay variation (CDV). This led to complex and confusing virtual circuit creation messages that were disliked by both vendors and users. It is also important to keep in mind that the work in the ATM Forum on QoS categories is an ongoing process that may be extended and refined with time. For example, extensions are being proposed for the best-effort category, UBR, to support an enhancement that will have minimum bandwidth guarantees. In order for ATM to provide high QoS for all anticipated services in the TM 4.0 specification, each of the following models would need to be supported [as].

Specification of acceptable QoS, such as cell loss, cell delay, and cell delay variation, is integral to cell relay service definition, as are other factors. Cell-loss guarantees vary with the selected QoS. Although some of the QoS parameters may be signaled to the degree of precision allowed by the length of the appropriate information element, a number of "packaged" QoS values may be available from carriers (e.g., Gold Service with CLR $\leq 10^{-10}$; Silver Service with CLR $\leq 10^{-9}$, etc.). Within the QoS parameters, carriers provide network assurance for

1. The sequential cell stream.

2. Cell integrity. Integrity includes correctness of the payload, assurance that the cell will be delivered, and assurance that a received cell is for the intended destination.

3. Low delay in establishing and releasing on-demand cells.

4. Low end-to-end delay for the delivery of the first user cell.

5. Low variation in intercell delay.

6. Traffic management (traffic control and/or congestion control). Traffic management controls efficient statistical multiplexing of user traffic. Traffic management depends on both the QoS selected/supported and the state of the network.

A final design consideration is whether the organization wants to develop a hybrid network that uses other fast packet services [e.g., frame relay and/or switched multimegabit data service (SMDS)] in some portions of the network.

9.3.3   WAN Design Issues

The three design issues listed at the beginning of this chapter have major applicability at the WAN level.

Topological design (when/where to use ATM technology?).   At this juncture, the most cost-effective use of ATM in WANs is for native-speed interconnection of islands of geographically dispersed LANs. Typically, ATM-ready local routers or Layer 2 switches (i.e., 10/100-Mbps Ethernet switches) can be interconnected via carrier-provided WAN ATM services. For small networks (e.g., a few backbone routers), the aggregation devices (e.g., routers) can be connected directly over permanent virtual connections (PVCs), say using Classical IP over ATM (RFC 1577) methods; when there are more than a few devices, then the use of switched virtual connections (SVCs) becomes a necessity. In addition to setup of VC channels on demand, SVC-based services include the use of LANE principles implemented over a WAN or at the central office as a way to provide multipoint connectivity, and/or MPOA principles. In general, however, today only PVC services are available from carriers. Wide area LANE services are not generally supported by carriers [user organizations can implement LANE privately using a carrier's permanent virtual path (PVP)—permanent virtual connection—tunnels, but this is not the same as the carrier's providing a multipoint SVC-based workgroup environment across a distance utilizing LANE]. Similarly, MPOA and L3S services that are integrated between the public and private networks are not yet available (see Chap. 7).

Performance (what are the end-to-end delay, jitter, and loss?).   This question relates to (1) the service class that may be utilized by the routers and (2) the kind of congestion control and overbooking policy utilized by the carrier's ATM hardware. Although this issue has caused a lot of confusion, the end user need not worry about how the carrier meets its stated grade of service. From the user's point of view, all that is needed is for the carrier to commit to support specified service-level agreements. The carrier needs to size the switch and provide options to support appropriate call admission control (CAC) disciplines and other traffic management mechanisms to meet the stated SLAs (see Sec. 9.3.4). What users can do is to inquire about the kind of oversubscription philosophy that the carrier employs; in general, users should look to carriers that do not oversubscribe beyond a 2-to-1 or at most a 3-to-1 level. However, the issue is somewhat complicated by the kind of services supported and also by the fact that CAC is calculated differently on different switches (there is no standardized recommendation on how to calculate it uniformly across all switch vendors). Section 9.3.5 provides an example of a CAC calculation that can be used (at the pragmatic level) to estimate the equivalent bandwidth that results on a link off a Fore switch, which in turn enables the carrier's capacity planner to size the appropriate interswitch link (the calculation off a Cisco switch would be different).

Capacity (when do new resources—switches and communication facilities—have to be added to meet the expected grade of service?).   This question relates to (1) the switch throughput, (2) the multiplexing strategy used in support of CAC, and (3) the addition of interswitch transmission facilities by the carrier.

9.3.4   Call Admission Control

ATM switches and IP routers that support QoS by policing their input traffic implement what is called the leaky bucket algorithm. Call admission control uses the traffic parameters [PCR, SCR, MBS, minimum cell rate (MCR), and CDVT] to make a best estimate as to whether the switch, or switches, can carry data between clients at the desired bandwidth. ATM QoS is specified to the software implementing the call admission control as the performance that can be realized by the end devices via the negotiation of the traffic description parameters.[2]

If the desired resources exist, then the switch will respond to the end devices that the call is established. The successful set of parameters is called the traffic contract because it specifies that the network will make some guarantees about performance if the network resources utilized are less than or equal to those specified in the call setup request. After the call has been established, the Usage Parameter Control software on the switch monitors the resources being used to determine if the traffic contract is being obeyed [as].

Often, selecting the values for the traffic parameters can be difficult because most applications, like LAN traffic, are not traditionally specified in terms of these metrics. In many cases network managers either specify no policing of LAN traffic or set the values of PCR/SCR only because their ATM carrier is selling limited bandwidth. In that case, the carrier can utilize ingress policing on the ATM ports serving a customer to check that only the allocated bandwidth is used [as].

For a virtual circuit requiring high QoS, or at least a very consistent stream of cells, the peak rate could be nearly identical to the bandwidth allocated. If the virtual circuit is carrying bursty data, the peak rate requested is typically somewhat lower than the actual maximum because the call is more likely to be accepted and the transmitting host's ATM Layer should be able to shape the egress cell stream.

The SCR and MBS are the parameters used in constructing a second leaky bucket measurement device. The SCR is always chosen by the end system when signaling for the creation of the virtual circuit and is by definition lower than the PCR. MCR is used only with the available bit rate traffic category described below. When considering SCR values, the network manager should consider values that will provide consistent good performance to applications over long periods of time. That way the data rate will be able to burst to the PCR for short periods but will always be able to depend on SCR performance [as].

The above traffic parameters are used by the network to determine which resources are required for the virtual circuit. In addition to those

---

[2]The traffic parameters are defined by the ATM Forum in the *User-Network Interface Specification*.

listed above, the following QoS parameters are signaled in a call setup message:

- Cell loss ratio
- Cell transfer delay
- Cell delay variation

These QoS parameters are used in the following fashion when requesting service (see Table 9.2):

- CLR applies to CBR, rt-VBR, and nrt-VBR.
- For ABR, a value of CLR may be associated with the service, but it is not signaled.
- CTD is carried in the call setup messages for CBR and rt-VBR services.

TABLE 9.2

ATM Service Category Attributes Defined in ATMF UNI 4.0

	ATM Service Category				
	CBR	**rt-VBR**	**nrt-VBR**	**UBR**	**ABR**
	Delay Sensitive		**Non–delay sensitive**		
Traffic Parameters					
PCR and CDVT[a,b]	Specified			Specified[c]	Specified[d]
SCR, MBS, CDVT[a,b]	N/A	Specified		N/A	
MCR[a]	N/A			N/A[e]	Specified
QoS Parameters					
Peak-to-peak CDV	Specified		Unspecified		
Max CTD	Specified		Unspecified		
CLR[a]	Specified			Unspecified	[f]
Other Attributes					
Feedback	Unspecified				Specified

[a]These parameters are either explicitly or implicitly specified for PVCs or SVCs.
[b]CDVT is not signaled. In general, CDVT need not have a unique value for a connection. Different values may apply at each interface along the path of a connection.
[c]May not be subject to CAC and UPC procedures.
[d]Represents the maximum rate at which the ABR source may ever send. The actual rate is subject to the control information.
[e]Work is under way in the ATM Forum to support MCR for VBR.
[f]CLR is low for sources that adjust cell flow in response to control information. Whether a quantitative value for CLR is specified is network-specific.

■ CDV is carried in the call setup messages for CBR and rt-VBR services.

The UNI 4.0 specification supports the additional features of [as]

■ *Traffic parameter negotiation.* This allows the SETUP to contain multiple information elements (IEs) for the same object with the intent of reducing CAC failures. If the first IE is unacceptable for call completion, the switch has the option of retrying the CAC with the second IE.

■ *Available bit rate.* This utilizes ATM's closed-loop flow control by requesting in the signaling message that the service be associated with the new circuit. ABR also functions as a means of traffic parameter negotiation because it allows the user to establish a baseline, then request modification after the circuit is in service and applications can learn from the network how much data can be transmitted per second.

■ *Multicast extensions.*
   ■ *Leaf initiated join (LIJ).* This enables clients to add themselves to a multicast tree without the root's knowledge.
   ■ *Anycast.* This allows one ATM address to correspond to multiple receivers.

■ *Virtual path switching.* This permits the signaling of an entire virtual path instead of the usual VPI/VCI granularity.

■ *Frame discard service.* This allows signaling to request that this service be associated with a circuit so that during congestion the partial packet discard algorithms are employed.

■ *Signaling of individual QoS parameters.* The UNI 4.0 specification diverges from the UNI 3.1 philosophy by selecting a QoS service category.

With these new features comes a set of new IEs that evoke and control their behavior. The new IEs used when establishing a virtual circuit on a UNI 4.0 capable switch are [as]

■ *Minimum acceptable ATM traffic descriptor.* This field is used with the ABR service category. It sets the baseline for the ABR service and specifies the lowest number of bits per second that can be transmitted. ABR setup parameters include the various objects used to initialize ABR:
   ■ Initial cell rate
   ■ Resource management (RM) round-trip time

- Data rate increment factor
- Data rate decrement factor
- Transient buffer exposure

- *Alternative ATM traffic descriptor.*   This field is used with the new feature of multiple IEs for the same parameter. If the ABR service is being selected, the alternative traffic descriptors are prohibited.

- *Selection of ATM traffic category.*   The ATM traffic category (CBR, nrt-VBR, etc.) is selected by a new field, transfer_capability, in the broadband bearer capability information element. This IE also specifies whether the signaling message is point to point or multicast.

- *Connection scope selection.*

- *Extended QoS parameters.*   These are used to specify the values of cell delay variation and cell loss ratios. The QoS parameters are not directly specified by the IETF.

- *End-to-end transit delay.*

9.3.5   Call Admission Control Example

This section describes the kind of configuration that is required by a network planner (say, working for a carrier supporting ATM). As noted, the CAC calculation is switch-dependent. The CAC calculation determines what the "equivalent capacity" is when multiple PVCs are mapped to a single outgoing link. The planner wants to map incoming bandwidth in such a manner that the oversubscription is not too severe; otherwise cell loss, delay, jitter, and throughput will suffer. CAC ultimately drives the number of buffers needed in the switch, and the size of the outgoing trunk.

For Fore, the equivalent capacity used in CAC is a complex equation that cannot be readily computed by the planner for multicustomer/multiapplication environments. Standard mathematical treatment of hard functions of this kind is to use a *convex* function, namely

$$\text{Equivalent capacity} \sim \alpha(\text{SCR}) + (1-\alpha)\text{PCR}$$

for $0 \leq \alpha \leq 1$.

Now, the Fore switch makes the assignment over a concentrated link such that

$$\sum_i \alpha[\text{SCR}(i)] + (1-\alpha)[\text{PCR}(i)] = 5\text{LR}$$

where LR = line rate.

The question is, how high can the PCR be allowed to go and still support QoS? As a simplification, assume that $PCR(i)=PCR^*$ and $SCR(i)=SCR^*$. Then, solving this equation, one has

$$n\alpha SCR^* + n(1-\alpha)\ PCR^* = 5LR$$

or

$$PCR^* = [5LR - n\alpha SCR^*]/[n(1-\alpha)]$$

where n is the number of PVCs. For $\alpha=0.5$, we have $PCR^*=(10LR-nSCR^*)/n$, so that

$$PCR = \min\ \{LR, (5LR - n\alpha SCR^*)/[n(1-\alpha)]\}$$

Note that if $n=1$, $PCR=\min[LR, (5LR-\alpha SCR^*)/(1-\alpha)]$. At $\alpha=0$ and $\alpha=1$, clearly this says $PCR=LR$.

Now assume that (for $n>1$) $SCR^*=LR/n$; then

$$PCR = \min\ \{LR, [5LR - n\alpha(LR/n)]/[n(1-\alpha)]\}$$

$$= \min[LR, (5-\alpha)LR/n(1-\alpha)]$$

Table 9.3 shows that this function is monotonically increasing, with the lowest value at $\alpha=0$. Hence, one concludes that $PCR=\min\ [LR, (5/n)LR]$.

It is important to note that for $n\leq5$, $PCR=LR$; for $n>5$, $PCR=(5/n)LR$. This in effect says that the switch can support 500 percent overbooking.

This calculation says that in order not to overflow the buffers on the switch, the maximum peak cell rate on an incoming link being mapped to an outgoing link with n users over it can be as high as LR when there are between one and five outgoing users, and $(5/n)LR$ when there are more than five users. In turn, this says that if the planner must support more than $(5/n)LR$ on each incoming link (with LR the rate of the multiplexed outgoing link), then the capacity of the outgoing link should be increased to LR', where LR' > LR. This is typical of the activity and calculations required by planners working at carriers providing ATM services. Another activity is the assignment of buffer pools. Buffer pool assignment (to the various service types) is now more an art than a science, in spite of the dozens of theoretical papers published on this issue.

9.3.6   Carrier Services

An increasing number of carriers now offer cell relay services. At the intra-LATA (local access transport area) level, Teleport Communications

TABLE 9.3

Calculation Supporting Overbooking

n	a	Factor	Min (LR, LR·factor)
10	0	0.5000	0.5000
10	0.1	0.5444	0.5444
10	0.2	0.6000	0.6000
10	0.3	0.6714	0.6714
10	0.4	0.7667	0.7667
10	0.5	0.9000	0.9000
10	0.6	1.1000	1.0
10	0.7	1.4333	1.0
10	0.8	2.1000	1.0
10	0.9	4.1000	1.0
10	0.99	40.100	1.0
100	0.1	0.05000	0.05000
100	0.2	0.05444	0.05444
100	0.3	0.06000	0.06000
100	0.4	0.06714	0.06714
100	0.5	0.09000	0.09000
100	0.6	0.11000	0.11000
100	0.7	0.14333	0.14333
100	0.8	0.21000	0.21000
100	0.9	0.41000	0.41000
100	0.99	4.0100	1.0

Group (TCG), Ameritech, Bell Atlantic, and Pacific Bell offer cell relay services at the DS3, OC-3, or OC-12 rates. At the inter-LATA level, TCG, Sprint, and Worldcom/MCI offer cell relay. Also see Chap. 8.

9.3.7 Private WAN Design

The design of *private* WAN cell relay networks entails similar considerations, but other factors must also be addressed. These include

- Equipment selection, based on required functionality
- Throughput requirements for the access links, the switches, and the trunking between switches

- The infrastructure topology, including hub (edge switch) location, major switch (core switch) locations, and interconnecting links required to ensure appropriate diversified routing
- Tariffs of communication facilities used (including ATM tariffs if public ATM is part of the hybrid design)
- End-to-end traffic flow requirements
- Diversity and reliability constraints (diversity groups, vendor diversity, link and site diversity)
- Performance constraints, including delay, delay variation, cell and PDU loss ratio.

In general, when designing a nationwide ATM network, switches that are less than 500 miles away can be cost-effectively connected with private-line trunking; if the distance is more than 500 it is cheaper to utilize a public ATM service and tunnel through it.

9.4 Design Tools

Users have seen an accelerated introduction of new networking technologies in the past 5 years. Unfortunately, usable network design tools and underlying design theory have not generally kept up with these developments. Such tools would assist users to discriminate between cost savings that the new services might possibly afford and simple vendor hype. Based on our experience, users tend to be punctilious in their pursuit of cost-effectiveness through unnecessary or marginally needed capabilities that appear to exhibit "great efficiency" in no domain other than the trade press or multicolor sales brochures.

The 1990s is the decade of broadband services, from megabits per second (such as switched multimegabit data service) to gigabits per second. Given the relatively high cost of these technologies compared to that of 9.6 to 28.8 kbps services, which have been the data communications staple for the past quarter of a century, it is critical that design methodologies be applied to demonstrate to any internal or external auditor that cost-effective use is indeed being made. Broadband affords economies of scale, particularly for large backbone applications—1 T1 is cheaper than 24 DS0s; 1 T3 is cheaper than 28 DS1s; 1 E3-c is cheaper than 3 E1s. However, the judicial use of analytical methods to design such networks remains critical.

Of course, most classical design techniques, models, and formulas are equally applicable to narrowband services. However, the cost implications are more critical for broadband. For example, the cost of a 1000-mi dedicated voice-grade line is about $900 per month; the cost of a same-distance T1 line is about $9000. If a manager installing voice-grade private lines miscalculates the number needed and installs, say, 11 instead of 10, that mistake would be a $10,000 a year mistake. If the manager installed 11 T1 lines instead of 10, that mistake would be a $100,000 a year mistake.

There has been some evolution in tools over the past decade. Initial network design tools focused on analog point-to-point and point-to-multipoint lines; starting in the mid to late 1980s, a number of tools added the capability of designing networks that include T1 lines and multiplexers. Now one is beginning to see the addition of fast packet capabilities; however, only a few of the high-end tools have added such capabilities.

A good design tool assists the planner to understand the traffic requirements and codify these correctly into the data structure to be employed by the tool. Considerations include the following: Is the traffic in support of a star, client/server, or peer-to-peer environment? What are the traffic volumes and the sources/sinks? Is the traffic bursty? What ancillary protocols must be supported, say, in addition to IP? This last item is important not only for interoperability considerations, but also because each protocol has traffic profiles characteristic to that protocol. In addition, if ATM is to be used to support video or voice, appropriate design principles must be brought to bear; video and voice networking requirements are considerably different from those for data (e.g., the traffic is isochronous; there may be a need for multicasting; error correction must typically be handled with forward error correction; the information cannot be buffered while in transit; the data rates may be higher; etc.).

As noted earlier, the issue of access is important. A good design tool will help the designer select the appropriate access vehicle, whether an LEC's fast packet service or a dedicated line (from an LEC or AAP) to the IXC. The access facility must be sized at the physical level; in addition, there may be other parameters that have to be sized—sustainable rate, the burst interval, etc.

For private ATM networks, the design tool will assist the planner to decide switch attributes. For example, does the user need a pure ATM switch, or is a multiservice platform better? What are the backplane/backbone speeds? What are the locations where the switches must be deployed? Do all switches have the same capabilities, or does the planner need some access (feeder) nodes and some core switching nodes? How does one design for reliability and diversified routing? How does one appropriately size up for traffic management (to meet desired service class/QoS goals)?

For hybrid ATM networks, some of the options that must be addressed by the design tool relate to the choice of dedicated long-haul facilities between switches or the use of an IXC ATM service, the sizing of the hybrid system for end-to-end QoS, and the provisions of mechanisms in support of diversified routing.

The following list establishes some of the requirements for WAN design tools:

- Support tariff-sensitive algorithms.
- Handle hardware details.
- Support various architectures to simplify the comparison process.
- Perform integrated access and backbone design.
- Perform integrated voice/data/video/image design.
- Be fast.
- Find an optimal solution.

A design tool enables the planner to (1) determine how well the existing network architecture can support increased traffic loads, (2) predict network performance before investing in a particular network technology, (3) perform quantitative cost-benefit analysis by evaluating potentially high costs of new technologies against improved network performance, and (4) optimize network performance by comparing alternative topologies and configurations.

When evaluating design tools, planners should examine a number of features. Among the important considerations related to design tool functionality are

- What computing platform is required (Pentium PC, Unix workstation, etc.)?
- Does the product support graphics?
- Does the tool use modeling or simulation?
- Does the tool incorporate tariff information?
- Does the tool incorporate a database of switch hardware, along with switch features?
- What is the performance of the tool (i.e., how fast does it complete a design)?
- Does the vendor provide support services (training, help desk, etc.)?
- How good is the documentation?
- Must the user purchase the system or can it be obtained on a license basis?
- What is the cost of the tool?

Planners can use the checklist of Table 9.4 to compare and/or select different design tools.

TABLE 9.4

Considerations
in Selecting
Design Tools

Simulation Method

  Discrete event (dynamic)

  Analytical modeling (mathematical)

Model Libraries

  Model suites

  Model completeness and accuracy

Model Parameters

  Access to model internals

Interfaces

  GUI

  Graphical network configuration

  Nongraphical API for network configuration

Presentation Capabilities

  Analysis tool

  Customization of display

  Animation

  Desktop publishing support

Simulation Efficiency

  CPU/processor

  Memory

  Disk

Licensing

  Floating license

  Node locked

Cost

  Initial

  Maintenance

  Extra modules

TABLE 9.5 Comparison of Design Tools Available at Press Time

Network Simulation Product	Vendor	Strengths	Weaknesses
Planning Workbench Suite	Bellcore	Group of network planning software for frame relay, ATM, and SONET; data sharing capabilities; high performance; traffic and survivability analysis.	Relatively expensive.
BONES	Comdisco	Good graphical interface. Event-driven. Similar to OPNET; signal flow model.	Low-level protocols.
CACI Network II.5 (comp and network) COMNET III (LAN, WAN, enterprise) L-NET (LAN performance) Simscript 2.5 (programming language)	CACI Products Co.	Easy interface. Supports a variety of network types (802.x, FDDI, etc.). Animated results. Does not require a high level of programming skill.	Not as flexible in terms of new protocols. Coarse statistics. Applicable only to small networks.
GPSS	Several vendors	Event-driven simulation language.	
NetMaker	Make Systems Inc.	Includes preconfigured router models (Wellfleet and Cisco). Supports a number of muxes and cross-connects. Modular architecture. Graphical user interface. Supports TCP/IP, XNS, IPX, AppleTalk; RIP; OSPF; T1/T3; Ethernets and FDDI.	Can be expensive (up to $130,000). No firm information on speed.
ObjecTime	BNR	Intended more for software design specification.	Slow.

Key design systems for contemporary enterprise networks include CACI, Cascade, Comdisco, MIL3, Mind, Quintessential, and WANDL. Not all of these focus on ATM. Based on the author's experience, CACI and MIL3 are currently the two major systems that support ATM technology in a prominent fashion. Table 9.5 compares features of a number of design tools available at press time.

9.5 Conclusion

In terms of design approach, organizations have a number of choices on how to proceed. They can do the design in-house without using any

TABLE 9.5   Comparison of Design Tools Available at Press Time (Continued)

Network Simulation Product	Vendor	Strengths	Weaknesses
OPNET	Mil 3 Inc.	Good graphical interface (advanced X-based). Hierarchical specification. Includes 300 communications and simulation primitives. Efficient simulation. Includes library of detailed protocol models (TCP/IP, ATM, X.25, etc.). Used by several high-end planners and researchers to study ATM. Finite-state machine $14,000 per license.	Slow for large problems. Users must be familiar with some C syntax.
SES Workbench	Scientific & Engineering Software		Not as complete a library as some other products.
Telecom Framework	Jade Simulations International	Specifically designed for "telcos"—SS7.	
TLC (Version 3.0)	Quad Design Technology	Transmission line simulator (ribbon, coax, backplanes).	
Network Planning and Analysis	WANDL	Supports WAN pricing. Fractional T1, T1, and T3 backbone network design capabilities. Node card port configuration management. Supports synchronization timing, access network, and voice design. Performs frame relay analysis.	ATM module not yet available.

design tools (we emphatically discourage this approach; however, it is often what happens); purchase commercial, off-the-shelf tools and use in-house design staff; outsource the design to a qualified firm; have the carrier assist with the design; or hire a system integrator. In the last two cases, however, it is likely that the user will not be afforded the optimal solution, since products and services offered by the vendor are likely to play a major role in the final recommendation.

9.6   References

[as] A. Schmidt and D. Minoli, *MPOA*, Prentice-Hall/Manning, Greenwich, Conn., 1998.

[dmdp] D. Minoli, *ATM Network Design*, Datapro Report, McGraw-Hill, June 1997.

[fo] Fore, Promotional Material, Courtesy R. Pybus.

Making Broadband Services Real: Carrier Challenges

10.1   Carrier Challenges

Carriers face a number of challenges in actually delivering ATM services to the user (see References, Sec. 10.7). Some of these challenges are as follows:

1. Identify which broadband services to actually deliver. These services have to have market potential and user interest. At the same time, these services have to be doable technically and make sense as WAN services. Naturally, the services have to be profitable for the carrier. The carrier needs to keep in mind that end users are not strictly interested in technology, but rather in capabilities that can enhance the user's enterprise networking posture.

2. Select an appropriate delivery architecture. This entails (*a*) a customer premises architecture, (*b*) an access architecture, (*c*) an edge/multiservice switch architecture, and (*d*) a core/tandem switch architecture.

3. Educate customers that WANs are not "flat." The network is not one small "cloud," but may interconnected clouds. The network may consist of networks from incumbent local exchange carriers (ILECs), competitive local exchange carriers (CLECs), interexchange carriers (IXCs), Internet service providers (ISPs), network service providers (NSPs), wireless providers, and international providers.

4. There must be a capability for rapid provision of services, since customers are not going to be willing to wait for long periods of time to get an order turned up. Rapid provision relies on systems such as order tracking, facilities tracking, equipment tracking, customer records, etc. Figure 10.1 depicts an example of the set of tools required by a typical carrier. The problem of rapid provision becomes more accentuated with the need to do multicarrier coordination.

5. Quality of Service (QoS) support is needed end to end not just for one but for the set of services under consideration, i.e., constant bit rate (CBR), variable bit rate (VBR), available bit rate (ABR), etc. This has several challenging facets, such as the selection of the correct ATM switch—namely, one that has good queue management, supports multiple priority levels, and has flow control, in addition to fundamental functions such as policing, tagging, and (possibly) shaping. End-to-end QoS is also dependent on the ability to do networkwide capacity planning. Furthermore, QoS may have to be maintained across administrative domains (i.e., across the multiple network "clouds" discussed above for LECs, IXCs, etc.). Neither P-NNI nor UNI-based carrier-to-carrier (or even WAN-switch-to-WAN-switch) is adequate for consistent and resilient end-to-end QoS management.

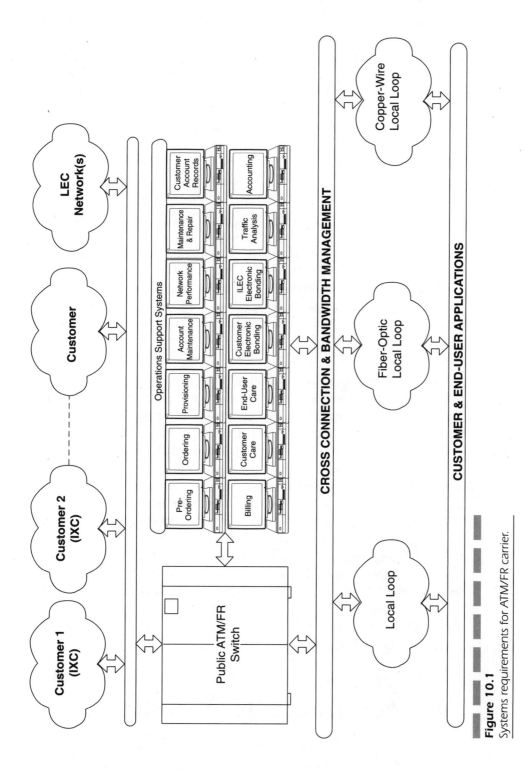

Figure 10.1

Systems requirements for ATM/FR carrier.

6. Reliability is a must, particularly when ATM is used for integration of all of a company's requirements (data, voice, video) over a single access link. This requires, at the very least, a redundant switch processor, redundant switching fabric, redundant line cards (optional), redundant power supplies, and redundant management links. Again, the delivery of high-availability, high-reliability services can be a challenge, particularly since many lower-end switches (e.g., 10 Gbps) are not carrier-grade and/or NEBS (National Equipment Builders Standards)–compliant. Yet, there is a desire to utilize smaller switches to facilitate entry into the market, rather than large (e.g., 50 Gbps) systems, at least in the short term and/or in Tier II or III cities.

7. Scalability is another challenge. A number of switches on the market operate at 1, 2.5, 5, or 10 Gbps (e.g., support 8, 12, 32, or 64 OC3 UNIs). Especially when starting at the lowest end, a design and method of procedure (MOP) is needed to scale up to higher capacity and port counts. This applies to Tier I cities and/or cities that may serve as a (nationwide) aggregation point (a typical nationwide ATM network may have 6 to 10 core ATM nodes).

8. Support of Network Layer functions, such as multiprotocol over ATM (MPOA), Internet routing, IP, etc., as discussed in Chap. 7, are required. There have been a number of proposals from vendors and standards bodies, but none has been really satisfactory. Meshed IP networks do not scale, whether on private lines or PVC-based ATM networks. Hence, an IP-supportive switched capability is required; however, services such as LAN emulation (LANE) and MPOA are really campus-based solutions and do not scale well to WAN environments; further, it is not clear what function the carrier provides, versus what the customer provides [e.g., LAN emulation configuration server (LECS), LAN emulation servers (LESs), MPOA servers, etc.]. It is worth noting that there is an abundance of new solutions at the LAN level, and very few at the WAN level. This is in spite of the fact that LANs are really cheap, and companies' communication expenditures (from a recurring charges point of view) are in the WAN; hence, one would hope to see creative efforts in the WAN, it being more onerous on budgets that the LAN.

10.2   Services to Be Offered

As already noted, in selecting services to be offered, it is important to distinguish between broadband at the desktop or campus level, for which there are many alternatives (e.g., switched Ethernet, 100-Mbps Ethernet, Gigabit Ethernet), and broadband at the metropolitan area network (MAN) or WAN, for which there are few if any alternatives. Carriers must be able to articulate this difference to the end users, and make the case that different goals, grades of service, economics, equipment, approaches, and business/regulatory environments exist in the two arenas.

Also, ATM is only a tool, not necessarily an end in itself. In fact, ATM is only a protocol—an agreement between Data Link Layer endpoints on how to send data (somewhat) reliably and effectively across a hop (link). The real goal is broadband connectivity supporting desired customer applications. These may include

- Wire-speed multipoint connectivity in the form of transparent LAN services, where the carrier provides a managed edge adaptation device that supports the customer legacy LAN, for example, 10/100-Mbps Ethernet, token ring, FDDI, and even Gigabit LAN extensions over a WAN at specifiable speeds (up to the maximum wire speed of the LAN)
- ATM UNI service at 1.5, $n \times 1.5$, 45, 155, and 622 Mbps
- Frame relay over ATM (network and service interworking, each because of intrinsic advantages that can be secured)
- Support of IP
- Voice over ATM
- Integrated access, specifically low-end data, Internet, voice, and/or video services for institutional users such as school districts

Figures 10.2 through 10.5 depict some examples of broadband services offered by Teleport Communications Group.

An issue related to services to be provided centers on the support of IP in the carrier's network, as discussed at the technical level in Chap. 7. From the carrier's perspective,

- LANE is a building/campus technology that uses bridging and is nonscalable.
- MPOA supports enterprise networks, but may well be impractical for carriers.
- In IP Switching/Tag Switching/MPLS, the question is, What is the carrier providing?
- Other questions are: How are end-system-to-end-system QoS in general (e.g., API-enabled communications) and Resource Reservation Protocol (RSVP) in particular supported? How is IP multicasting supported?

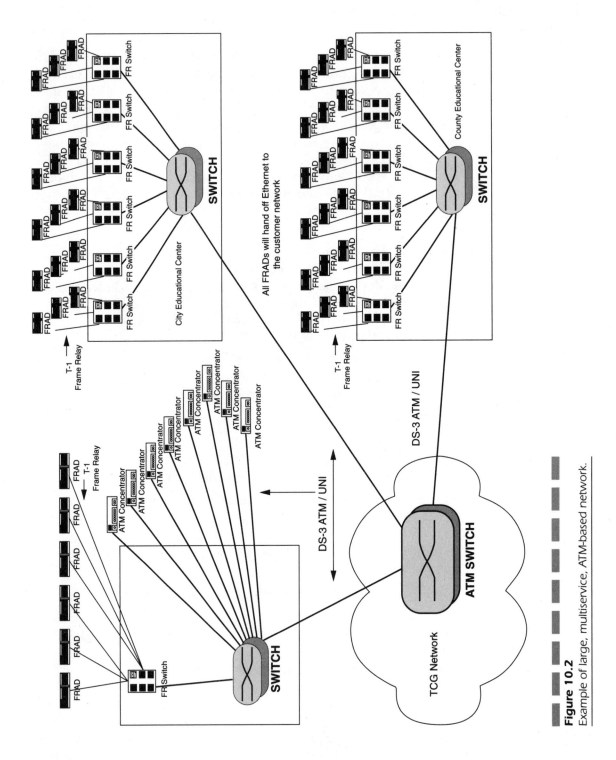

Figure 10.2
Example of large, multiservice, ATM-based network.

Figure 10.3
Example of ATM
application.

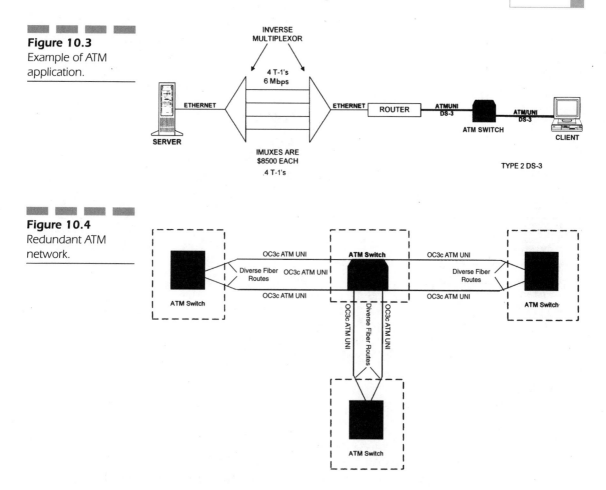

Figure 10.4
Redundant ATM
network.

Another set of questions related to service to be provided has to do with the customer's own architecture: Does the customer use routers in the enterprise network with switched Ethernet to the desktop? Does the customer use campus LANE or MPOA? Is Tag Switching/MPLS utilized?

10.3 Switch Requirements

As noted earlier, the carrier must select the right switch in order to be able to support broadband services in an effective manner. There is likely to be a multitiered hierarchy of switches. Therefore, the first step in any

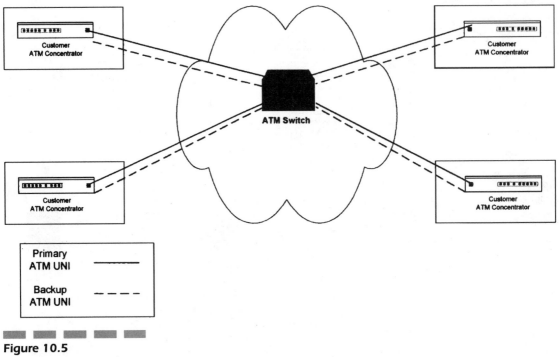

Figure 10.5
ATM network—example.

service rollout is to decide what the architecture will be. Figure 10.6 depicts one example of an architecture. The figure shows TCG's architecture, which is composed of managed customer premises equipment (CPE), legacy-to-ATM interworking equipment, multiservice edge ATM switches, and core ATM switches (e.g., to handle tandeming off to an IXC, or be a handoff to the company's own long-haul network). In that architecture, the carrier may well reach the conclusion that the switches on the market are too small. For example, if the carrier anticipates 1000 OC-3 ATM customers in Tier I cities such as New York, Boston, and Chicago, then a 150-Gbps switch may be required. However, as noted, some carriers may choose to start out smaller and then grow as needed.

Fortunately, the practical observation related to ATM, particularly for permanent virtual connection (PVC) service but also for switched virtual connection (SVC) services, is that generally there is no real any-to-any connectivity requirement across the plethora of inlets, as might be the case in a voice switch. In a voice environment, theoretically any inlet could want to connect to any outlet. This usually requires a multistage matrix

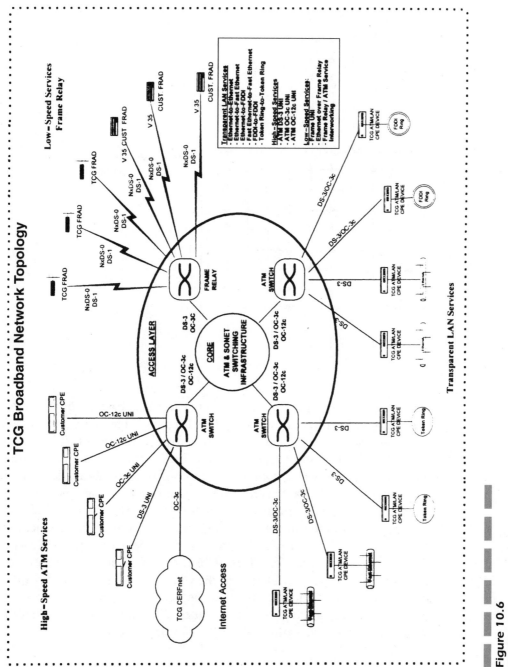

Figure 10.6

TCG broadband network topology.

Figure 10.7
Growth mechanism
in voice switches ver-
sus growth mecha-
nism in ATM
switches.

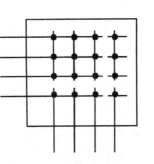

Voice switch with 10 inlets and 10 outlets (100 crosspoints)

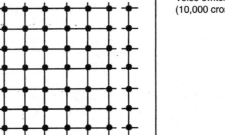

Voice switch with 100 inlets and 100 outlets
(10,000 crosspoints), single stage

(a)

that facilitates the connection without having to have N-squared crosspoints (whether physical or just cell registers) (see Fig. 10.7a). Also, should
the switch start out with an $(N/j) \times (N/j)$ matrix, where N is the ultimate
population and j is some scaling divisor, then as the requirement grows to
$2 \times (N/j)$ inlets/outlets, a complex three-stage routing matrix is typically
required (see Fig. 10.7b). This may well entail two first-stage modules, one
middle-stage module, and three output-stage modules. In data, the ports
are usually restricted in terms of potential connectivity requirements to
a few "localized" closed user groups. For example, a 16-port ATM switch
may terminate 3 ports for Citibank, 6 ports for Chase, and 7 ports for
First Boston. It is very unlikely that the Chase ports need to set up a PVC
or an SVC to the Citibank ports. As the need for more ports grows, the
planner can simply add first-stage modules, without the need for a middle-
stage module, which would be needed for any-to-any connectivity; if a
very few ports need to cross a module, then a corresponding few inter-
module links can be established (without the full middle-stage apparatus).
See Fig. 10.7c. For example, there might be a need to add, say, 7 ports for

Figure 10.7
(Continued)

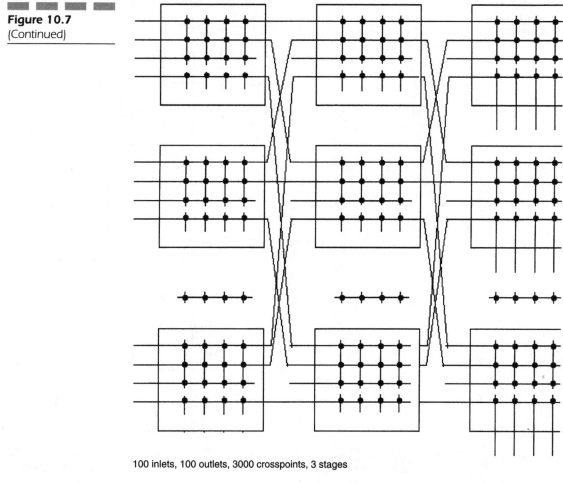

100 inlets, 100 outlets, 3000 crosspoints, 3 stages

(b)

First Boston. These ports would terminate on the second module. Now if Citibank needed an additional interconnected port, the new port could be added to the second module, and Citibank-wide connectivity would be achieved via an interswitch trunk. As noted, this tends to be the exception rather than the rule. The net effect is that ATM switches (say of 10 Gbps) can be added "linearly" (i.e., without complex middle stages) as the need arises.

Hardware redundancy is a must, as is port density. Some of the switches on the market only have one, two, or four ports on a card (particularly at the DS3 and OC-3 rates); although some manufacturers support up to

Figure 10.7
(Continued)

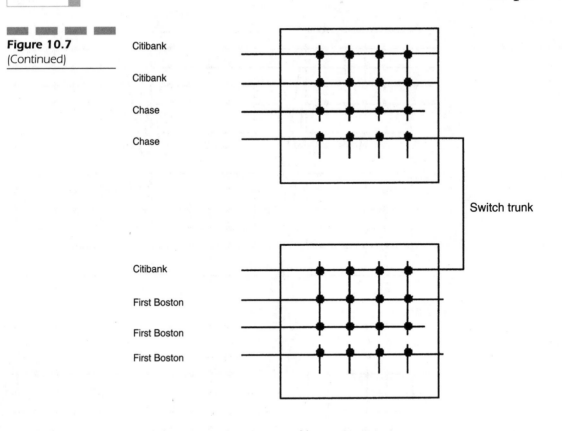

Citibank

Citibank

Chase

Chase

Switch trunk

Citibank

First Boston

First Boston

First Boston

(c)

eight ports per card, this is still relatively small. Low port density translates into rack space and large footprints in the central office, where space is always at a premium. Edge switches typically have to support multiprotocol adaptation—for example, frame relay–to–ATM interworking (this is in addition to the CPE adaptation equipment, if needed).

Good buffer management is needed to achieve sustainable end-to-end QoS-based multicustomer, multicarrier, multiprotocol support, as discussed in Chap. 9. End-to-end also means across administrative domains.

There is a critical need for capacity planning mechanisms and tools, again end-to-end across the network. Effective capacity planning is needed, in conjunction with good buffer management, to achieve end-to-end QoS support.

Effective network management tools are also needed. In general, most switches have their own Network Management Systems (NMSs), which provide a reasonable good level of Element Manager functionality. The problem is that a carrier usually has several different nodes (e.g., it could have two switches from two different vendors deployed for competitive reasons; similarly, it could have two vendors' CPE). Therefore the desire is not to have to require the network operations center (NOC) technicians to learn multiple NMSs. Therefore, the goal is generally to operate in an environment such as HP OpenView (HPOV). The problem with this approach, however, is that the customized functionality (especially the GUI-based functionality of the NMS) may not all be available under HPOV. This then requires the technicians to learn the command-line interface (CLI) specific to each NMS. There are three goals in this arena:

1. Do not require the use of Ph.Ds in the NOC to run the network.

2. Be able to support multivendor functionality down at the detailed level, using a kind of manager of managers (MOM) approach, so that a single craft interface is visible to the technicians.

3. Have the ATM equipment support the telecommunications management network (TMN) standards.

Another requirement faced by carriers is to provide customer network management (CNM) capabilities. Users are interested in monitoring performance and outages, and perhaps being able to do some reconfiguration (e.g., add/change/delete PVCs). Many ATM tariffs are based on traffic contracts, hence users want to know if they are actually sending data at the sustainable cell rate (SCR), and what is the peak cell rate/maximum burst size (PCR/MBS) treatment on the part of the carrier; also, there may be interest in other performance metrics discussed elsewhere in this book. At the fault management level, there may be interest on the part of the end user in receiving Simple Network Management Protocol (SNMP) traps from the network. Configuration is typically a "stretch goal" because carriers do not want users to change the traffic contract or modify PVCs, since the monthly recurring charge is based on these parameters. Subtending all of this, there is the need for access security as related to the ability of seeing/modifying somebody else's network. A number of carriers, notably TCG, have developed Web-based interfaces to support CNM functions.

Another requirement for switches is the support of effective interconnection for carrier-grade communication. P-NNI is inadequate for carriers, since it was developed specifically for campus environments. Unfortunately, very few switches currently support B-ICI (Version 2).

There also are philosophical considerations related to the switch selection. Some carriers started out with a private line service (say, over

SONET rings); then they might have added frame relay services; and then they might have added cell relay services. Other carriers, however, may have taken a different course. They may have gone from providing private line services to providing ATM services; afterward, they may plan to add frame relay. Clearly, the sequence will affect the kind of technology that is selected. For example, a carrier that has deployed an ATM backbone may enter the frame relay market by either

- Deploying edge multiprotocol frame relay switches that handle frame relay as a service interworking with ATM (i.e., that support frame relay ports on the UNI but relay on ATM NNIs into an ATM cloud for actual remote switching). One example of a switch that can do this is the Ascend switch discussed in Chap. 6; the ATM switch is whatever switch has been utilized to build the ATM cloud.

- Augmenting the ATM switch with frame relay modules, if the switch supports frame relay directly on the chassis. One example of a switch that can do this is the ECI Telematics switch discussed in Chap. 5. At press time there were not many switches that in fact support frame relay and ATM in the same chassis, although several have announced their intention to do so in the future.

The actual selection of a vendor in either the frame relay or ATM market is somewhat complicated by the fact that several choices may be available, given the maturity of the frame relay and/or ATM technology. A taxonomy of switch vendors could include the following:

- Established vendors and new entrants
- Large vendors and boutique vendors
- Competitive but sustainable vendors and "buy-the-business" vendors
- Incumbents and new vendors

Table 10.1 and Fig. 10.8 depict a scoring matrix and some points for a set of switches recently evaluated by one of the authors.

In the area of near-term requirements (late 1990s), carriers (such as TCG) may require functionality such as

- Inverse multiplexed ATM (IMA) support (in both switch and CPE)
- Digital subscriber loop access multiplexer (DSLAM) functions for interworking with xDSL access technologies

TABLE 10.1 Sample Switch Scoring Matrix

Criteria with Rating System	Vendor A	Vendor S	Vendor E	Vendor N	Vendor C	Vendor B
1. Mandatory technical features (35)	32	29	24	30	30	27
▪ Feature interworking (1–3)	3	3	3	2	3	3
▪ Feature traffic management (1–3)	3	3	3	3	3	3
▪ Feature priorities (1–3)	3	3	3	3	2	3
▪ Feature other (1–3)	3	2	1	2	3	2
▪ Additional useful technical features (1–5)	5	5	3	4	5	3
▪ Multiservice platform (1–5)	3	4	3	4	4	2
▪ Conformance to standards (1–5)	4	5	2	4	4	3
▪ Physical requirements (1–8)	8	4	6	8	6	8
2. Reliability and scalability (35)	28	17	15	28	33	26
▪ Element redundancy (non-I/O) (1–5)	5	1	2	5	5	5
▪ Full gamut of high-density FR/ATM trunk plug-in (1–10)	10	10	3	8	9	8
▪ Additional non-FR user side plug-in (1–5)	1	3	5	4	5	3
▪ Local service scalability/topography (space # based on phase 3 #s) (1–5)	5	1	2	4	5	2
▪ Macrolevel LD-supporting ATM infrastructure scalability (1–10)	7	2	3	7	9	8
3. Network management (30)	28	18	21	28	24	17
▪ CNM/reports (1–10)	10	5	7	10	8	7
▪ Cohesive NMS architecture (1–10)	10	8	9	8	8	5
▪ NMS feature (fault, configuration, security, accounting, and performance) (1–10)	8	5	5	10	8	5

TABLE 10.1 Sample Switch Scoring Matrix (Continued)

Criteria with Rating System	Vendor A	Vendor S	Vendor E	Vendor N	Vendor C	Vendor B
4. Vendor strength and support (30)	29	7	23	26	30	29
■ Maturity and penetration (1–5)	5	1	4	5	5	5
■ Carrier-embedded base (1–5)	5	1	4	4	5	5
■ U.S. carrier/CAP base (1–5)	5	1	2	3	5	5
■ Nationwide presence operations/support (1–10)	9	2	8	9	10	10
■ Ease of integration with TCG base (1–5)	5	2	5	5	5	4
Subtotal	117	71	83	112	117	99
5. Pricing and terms (35)	27	28	28	16	20	22
■ Entry-level pricing/price per port (1–4)	3	3	4	1	2	2
■ Total cost of phase 1 (1–7)	4	3	7	1	4	5
■ Total cost of phase 2 (1–7)	6	6	7	3	4	3
■ Total cost phase 2+1000 ports (1–7)	7	6	5	3	5	3
■ Terms and conditions (1–10)	7	10	5	8	5	9
Total (165)	144	99	111	128	137	121

Figure 10.8

Analytical comparison
of actual switching
products performed
by authors.

	Technical	Scaleability	Net Manag	Vend Strngth	Pricing
Vendor A	91.43	80.00	93.33	96.67	77.14
Vendor C	85.71	94.29	80.00	100.00	57.14
Vendor N	85.71	80.00	93.33	86.67	45.71
Vendor B	77.14	74.29	56.67	96.67	62.86
Vendor E	68.57	42.86	70.00	76.67	80.00
Vendor S	82.86	48.57	60.00	23.33	80.00

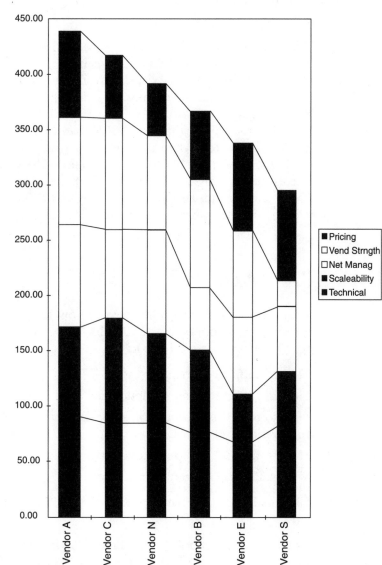

- Layer 3 support in the switch (e.g., IP Switching, Tag Switching, MPLS)
- High port density on switches
- Voice processing support, including decompression (if compressed) and handoff to central office voice switches (e.g., Lucent 5 ESS or Nortel DMS 500), including proper signaling support
- Wireless ATM support
- Closed-loop congestion control

Support for OC-48 and OC-192 ATM is also of interest for the core switch, edge switches, and CPE. This can support the "endogenous" use of ATM internal to the carrier's network, to support all of its services, including traditional telephony. In this application, ATM serves as a SONET manager and "overbooks" the SONET bandwidth by, for example, "telling" the voice switch (e.g., 5 ESS) that it has, say, 10 OC-3 trunks to another voice switch, when in reality the voice switch is connected via those 10 OC-3 ports to 10 OC-3 ports on an ATM switch, which in turn has access to, say, only 6 OC-3 trunks. In this way, the ATM switch "fools" the voice switch into thinking it actually has the 10 trunks. This drives efficiencies in the transmission plant.

10.4   Access Issues

Key consideration needs to be given to how to connect customers cost-effectively when supporting $n \times T1$, 10 Mbps, xDSL, and/or unbundled loops. Other access issues include:

- How to support Internet access
- How to integrate wireless support for ATM broadband
- How to make use of common space for multitenant/multiresidence buildings
- How to support high-scale residential services
- What CPE/edge concentrator to use

CLECs such as TCG have had a need for the following kind of ATM-based CPE:

- Devices able to support (via different plug-ins) Ethernet, Fast Ethernet, token ring, FDDI, CE, etc., for less than $5000 a device
- Support of DS3 and OC-3 ATM on the uplink, including long-reach optics

- Support of ATM inverse multiplexing and xDSL on the network and/or user side
- Blade directly insertable into the SONET add/drop multiplexer (desirable)
- Support of wireless trunks (e.g., 38-GHz radio)
- Low-cost plain old telephone service (POTS)+ATM
- Support of congestion control mechanisms
- Support of high-end residential services, since there is little hope for FTTCab, FTTC, FTTB, FTTH technologies at this time

Table 10.2 and Fig. 10.9 depict an evaluation of a number of CPE products (names removed) to illustrate the state of the art at press time and the criteria for selection.

10.5   Service-Supporting Carrier Processes

As noted, a plethora of effective processes and supportive systems are required in order to deliver broadband services in an effective and timely manner. Some of the key processes are

- Order tracking (with more than 200 variables)
- Facilities tracking
- Equipment tracking
- Configuration tracking
- Traffic contracts tracking
- Billing
- Trouble reporting
- Capacity tracking
- CNM interactions
- TMN

These are critical, challenging issues. As an example, one author at TCG has spent 8 staff-years selecting technology (e.g., multiservice edge switch, core switch, CPE) and 32 staff-years developing back-office support systems and processes.

TABLE 10.2

Sample CPE Scoring Matrix

Criteria with Rating System	Vendor X	Vendor S	Vendor L	Vendor C	Vendor A
1. Mandatory requirements (50)	47	39	44	45	36
■ Physical requirements	6	5	5	6	4
■ LAN interfaces	20	15	20	20	10
■ WAN interfaces	21	19	19	19	22
2. Mandatory technical features (50)	60	56	60	60	36
■ Bridging	13	12	12	13	9
■ Routing	7	0	5	9	2
■ STDs compliance	6	5	4	5	4
■ Closed user groups	3	2	2	3	2
■ Performance	12	18	17	5	3
■ Network management	19	19	20	25	16
3. Vendor strength and support (20)	15	10	12	24	21
■ 24×7 tech support	4	4	4	5	4
■ Maturity and penetration	3	2	2	5	4
■ Carrier-embedded base	2	1	2	4	5
■ Nationwide presence and support	3	1	2	5	4
■ Vendor's business outlook	3	2	2	5	4
4. Pricing and terms (50)	40	40	31	6	28
■ Entry-level pricing	30	30	25	5	28
■ Price for additional ports	10	10	6	1	N/A
■ Terms and conditions	0	0	0	0	0
Total	162	145	147	135	121

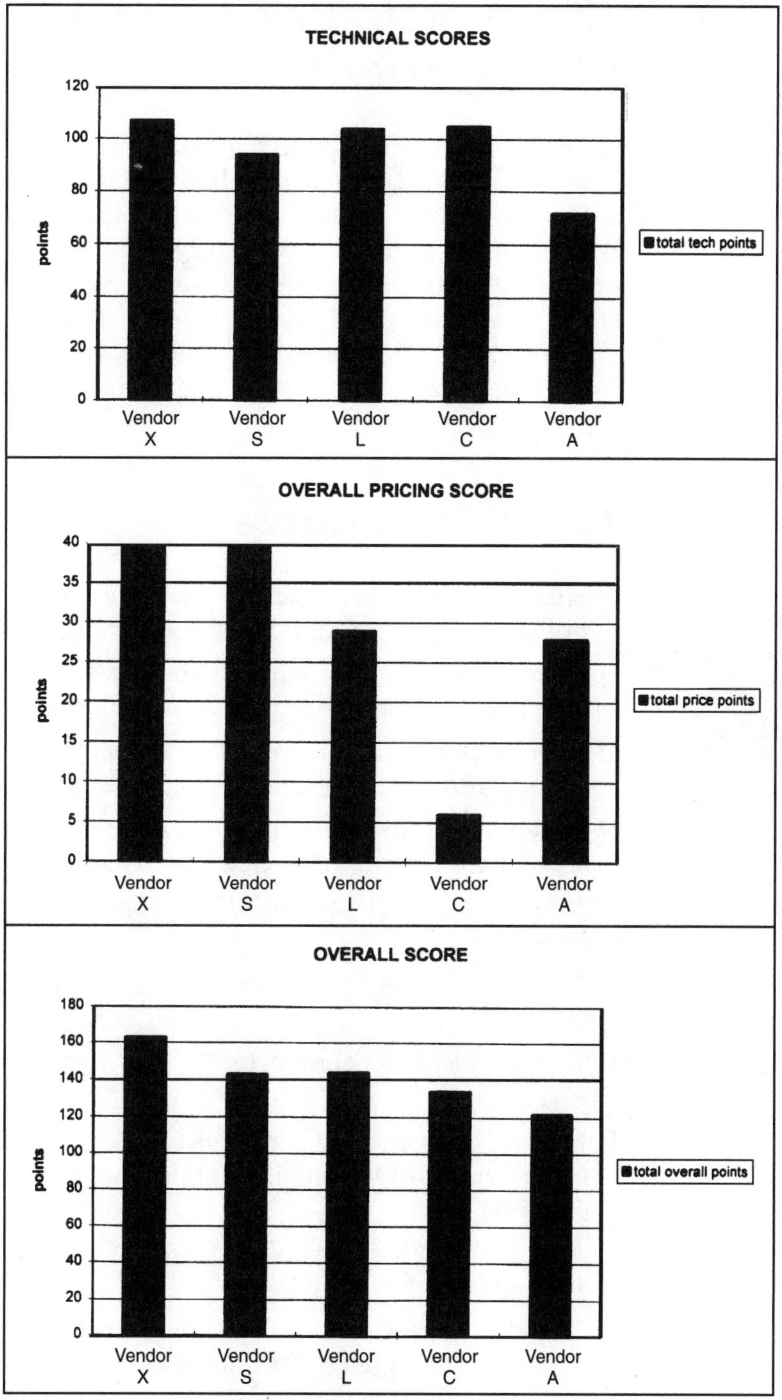

Figure 10.9
CPE comparisons.

TECHNICAL SCORES

OVERALL PRICING SCORE

OVERALL SCORE

10.6 Customer Network Management

CNM is a critical requirement for advanced cell-based networks. In the analyses undertaken by one author, a weight of about 20 percent was assigned to this feature. Fault and performance management are required as a minimum; configuration management may also be of interest. Figure 10.10 depicts a prototype TCG home page for TCG's Web-based service. Unfortunately, only about half the frame relay switches on the market support CNM, and only about one-quarter of the ATM switches on the market at press time had meaningful development under way.

10.7 References

D. Minoli, "Service Level Agreements for Outsourced Networks," The Information Technology Outsourcing Institute, New York, February 1995.

D. Minoli, "ATM Opportunities for the Futures Industry," Futures Industry Association Meeting, New York, April 1995.

D. Minoli, "Designing End-to-End Networks for New Multimedia Applications," ICA, Portland, Ore., April 1995.

D. Minoli, "Analyzing Outsourcing Strategies," Developing an Outsourcing Strategy Conference, Monterey, Calif., May 1995.

D. Minoli, "An Overview of ATM," Nortel's Forum, Toronto, Canada, June 1995.

D. Minoli, "ATM: A Status Review," TechCon/NeoCon, Chicago, June 1995.

D. Minoli, "Next Generation LANs," TechCon/NeoCon, Chicago, June 1995.

D. Minoli, "Outsourcing Communications in the Health Care Industry," Conference on Outsourcing Support Services in Hospitals, Chicago, July 1995.

D. Minoli, "ATM: Opportunities for End-Users," CMA Show, New York, September 1995.

D. Minoli, "ATM: Carriers' Challenges in Supporting Fastpacket and ATM Services," Interop 1995, Atlanta, September 1995.

D. Minoli, "Outsourcing Telecommunication Functions to Carriers," Systems Support Expo, San Francisco, October 1995.

Data Services LOB

TCG HOME PAGE

TCG CERFnet INTERNET SERVICES HOME PAGE

TCG is among the nation's leading voice and data communications services. Our business is focused in these key areas: managed network services, data communications products, and microwave communications. We have experience in implementing complex data networks on a global basis.

We develop and supply a wide range of digital access, bandwidth management, and internetworking products to meet the growing need to link local and wide area networks.

In the area of network services, TCG operates a number of different ones covering the healthcare data network, banking/financial services, chemicals, petroleum, entertainment, computers/electronics, government, and education. In the healthcare arena, we built the ATM communications infrastructure for the Staten Island Hospital and Epascabole Health Hospital.

We are the market leader for OC-3, OC-12, ATM, LAN interconnect, and microwave-satellite-based communications and one of the largest and best-known service professionals in data communications networks.

Network Management Report

Company Name: ☐ ☐ Enter ☐ Reset ☐

Figure 10.10
Prototype TCG CNM page.

D. Minoli, "An ATM Tutorial," 1995 Broadband and Multimedia Comfo-rum, Miami, December 1995.

D. Minoli, "Tutorial on ATM LANE and Classical IP over ATM," CMA Winter Session, New York, February 1996.

D. Minoli, "Emerging Technology and High Speed Data Networking," TCG Industry Seminar, Boston, April 11, 1996.

D. Minoli, "Voice, Video, and Multimedia over ATM," CMA Spring Session, New York, June 1996.

D. Minoli, "ATM-based Telemedicine Applications," 1996 Global Telemedicine and Federal Technologies, Williamsburg, Va., July 1996.

D. Minoli, "Emerging Technology and High Speed Data Networking," TCG Industry Seminar, Chicago, September 1996.

D. Minoli, "ATM and Broadband," 1996 First Annual Telecommunications Conference, Austin, Tex., October 1996.

D. Minoli, "Keynote Speech: Network Issues in Convergence," IEEE Convergence of Networks and Services, New York, November 1996.

D. Minoli, "ATM in Supercomputing Environments," Supercomputing 1996, Pittsburgh, Pa., November 1996.

D. Minoli, "Voice over ATM, Making It a Reality," 1996 Integrated OSS for ATM Broadband Technologies, Orlando, Fla., December 1996.

D. Minoli, "Convergence of Networking Technologies," 1997 Conference on Emerging Technologies, Nashville, Tenn., January 1997.

D. Minoli, "Making Broadband Services Real," 1997 Interop, Las Vegas, Nev., May 1997.

D. Minoli, "High-Performance Networking via IP Switching," Baltimore ITEC Expo, May 1997.

D. Minoli, "Carrier's Challenges in ATM Delivery," 1997 Comdex, Atlanta, June 1997.

D. Minoli, "IP Switching, Tag Switching, MPOA, MPLS," Nordex Wall Street Seminar, New York, June 1997.

D. Minoli, "Selection of Carrier Class ATM and Frame Relay Switches," The Annual ECI Telecom Technical and Planning Symposium, Washington, D.C., July 1997.

D. Minoli, "Selecting a Carrier for High-End ATM Services," International Engineering Symposium, Chicago, August 1997.

D. Minoli, "Carrier's Roles in Making ATM Real," 1997 Broadband Networking Symposium, Washington, D.C., November 1997.

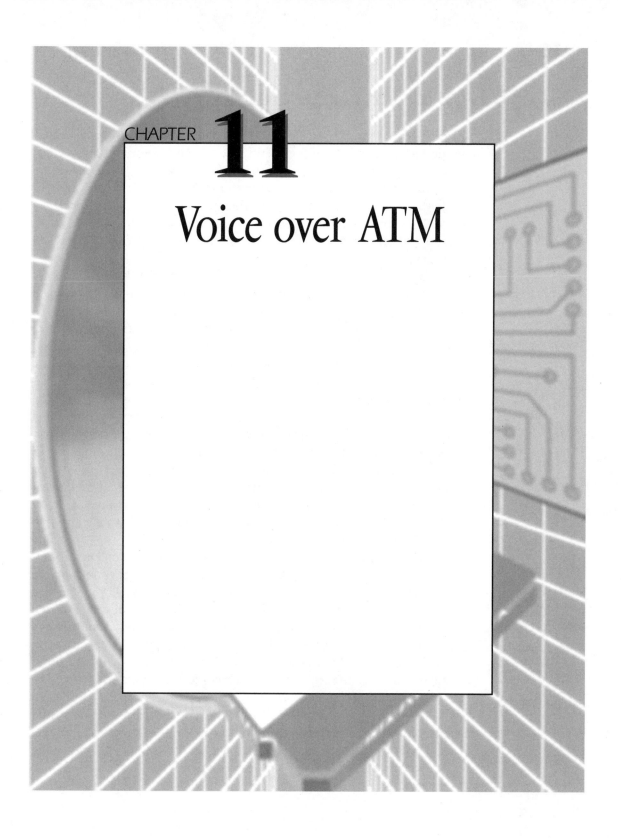

CHAPTER

11

Voice over ATM

11.1   Introduction

Asynchronous transfer mode was developed to be a multimedia, multi-service, multipoint technology. Until the present, however, the practical focus of developers of carriers' equipment, of developers of users' equipment, and of service providers has been on the data side, e.g., IP support. Major activities, such as the ATM Forum's user-to-network specifications, LAN emulation (LANE), and multiprotocols over ATM (MPOA), have implicitly concentrated on data applications, whether at the Data Link Layer or at the Network Layer.

Because of all the flurries of activities in the enterprise networking industry in the 1990s, such as the rolling out of frame relay service, deployment of upgraded desktop and backbone enterprise networks, expanded use of the Internet, introduction of intranets, and corporate deployment of ATM, the question of whether it makes economic sense to "carry voice over data networks" naturally arises. Three variants are possible: (1) voice over frame relay, (2) voice over ATM, and (3) voice directly over IP (i.e., over routed enterprise networks and the Internet).

Since voice technology has been around for over a century and digital voice has been deployed since the early 1960s, the real questions for corporate network planners are: (1) Is voice over ATM a practical reality, or just a technical possibility? (2) How economic is voice over ATM? (3) Is there customer premises equipment available to support voice over ATM? (4) Can a standard PBX be supported? (5) What is the quality of voice? (6) Are there standards for voice over ATM? Similar questions apply to voice over frame relay (over ATM).

Proponents proclaimed 1997 as the "Year for Voice over Data." Since then, voice over packet networks, including the Internet, has continued to see penetration. Equipment supporting voice, specifically voice-enabled frame relay access devices (FRADs), is now available from over a dozen vendors, and voice over frame relay technology is seeing commercial introduction. The Frame Relay Forum recently approved a specification to enable multivendor interoperability. Voice over ATM, on the other hand, is still in its infancy and is just a technical novelty at this time. The economics are not yet favorable, unless an organization already has an ATM-based enterprise network that uses ATM via edge multiplexers, and the organization is interested only in on-net voice. Some carriers, in contrast, are looking to use ATM totally internal to their network as a statistical multiplexing technology to derive higher SONET efficiencies. Such an arrangement would support voice in a new way that is fundamentally different from today's synchronous networks. However, significant roll-out of this radical architecture is not expected any time soon.

This chapter discusses the positioning of voice over ATM and voice over frame relay (since frame relay is now often delivered over ATM). It provides a review of compression and integration methods, and some vendor and market information for both ATM and frame relay [dmdp1] [dmdp2].

11.2   Voice Support in ATM Environments

11.2.1   Evolving Voice Applications and Directions

In recent years there has been a movement toward the use of packet technologies, such as frame relay and ATM, in corporate enterprise networks in order to achieve the all-points broadband connectivity that was discussed in Chap. 7 in a cost-effective manner. However, some packet technologies that may be part of these networks may not support voice-grade Quality of Service (QoS), e.g., IP.

ATM has three intrinsic advantages over other networking technologies when it comes to voice:

1. ATM was designed from the start to be a multimedia, multiservice technology. The very format of the cell was arrived at by considering data, voice, and video payload requirements.

2. ATM supports extensive QoS and service class capabilities. This allows time-sensitive traffic, such as voice, to be transported across the network in a reliable, jitter-free manner. Service classes are supported, in part, by various ATM Adaptation Layers (AALs), discussed in the next section.

3. Switches have been designed with effective traffic management capabilities (e.g., call admission control, Usage Parameter Control, traffic shaping, cell tagging, cell discard, and per-VC queue management) to support the Quality of Service and service classes needed for the various applications, including voice.

These capabilities have already been put to good use in data applications. At present it seems that ATM to the desktop will not be popular until at least the turn of the century. This is because there are multiple competing technologies, some of which require only minimal infrastructure upgrade to be deployed. In particular, switched 100-Mbps Ethernet is well positioned to cover this space. However, when it comes to broadband wide area network (WAN) applications, ATM is the only technology available at this time. Hence, ATM is seeing deployment in this space, either as router-to-router technology or as campus switch–to–campus switch technology. The service is secured either via an organization-built WAN or via a carrier's public network.

As just implied, many *Fortune* 500 companies now use ATM services. Hence, there is interest in addressing the question of services and media integration. See Fig. 11.1 [em]. Integration has proved to be reasonably effective in the frame relay context for the support of small office/home office (SoHo) locations. However, the pertinent questions for corporate network planners were identified earlier and focus on economics, standards and interoperability, product availability, and PBX and central office voice switch support. Given this context, it should be clear why the switch requirements described in the previous chapter called for both CPE and edge/core ATM switch support of voice as well as 5ESS/DMS connectivity. Table 11.1 depicts some of the key requirements for the support of voice, which any new alternative architecture is expected to accommodate. The bottom line at this juncture is that voice over ATM is still in its infancy and is a technical novelty. Some carriers, including Teleport Communications Group, have demonstrated the feasibility of network-supported ATM voice; however, existing alternatives make voice over ATM not the least-cost solution.

Business voice requirements continue to be key for the overwhelming majority of businesses. There are several "traditional" and "untraditional" alternatives to the support of voice. On the traditional side, carriers such as AT&T, Sprint, and Worldcom/MCI now have bulk-rate

Figure 11.1
Voice over
ATM: advantages.

Allows voice and data to share bandwidth—physical resource sharing.

Allows for a common network management approach.

Supports voice compression and silence suppression, which increases the number of simultaneous calls that can be handled. Savings on the order of 16:1.

Supports fax relay for efficient use of bandwidth. Savings on the order of 8:1.

TABLE 11.1   Basic Voice-Feature Requirements for Voice over Data Applications

Feature	Description	Requirement in ATM	Requirement in IP Net	Requirement in Frame Relay
Compression	Sub-pulse code modulation (PCM) compression significantly reduces the amount of band-width used by a voice conversation, while maintaining high quality.	Nice to have.	Must have.	Must have.
Silence suppression	The ability to recover bandwidth during periods of silence in a conversation makes that bandwidth available for other users of the network.	Nice to have.	Must have.	Must have.
QoS	Assuring priority for voice transmission is critical. This keeps delay, delay variation, and loss to a tolerable minimum.	Must have. ATM has been developed with significant QoS/traffic management support.	Must have. Very little current support [type of service (TOS) is not generally implemented in routers]. There is a hope that the Resource Reservation Protocol (RSVP), which reserves resources across the network, will help. However, RSVP is only a protocol; intrinsic network bandwidth must be provided before a reservation can be made.	Must have. Frame relay does explicitly support priority-based QoS. Recent development is attempting to address this limitation.
Signaling for voice traffic	Support of traditional PBXs and the associated signaling is critical.	Must have for real applications.	Must have for real applications.	Must have for real applications.
Echo control	Echo is annoying and disruptive. Control is key.	Must have for real applications.	Must have for real applications.	Must have for real applications.

TABLE 11.1 Basic Voice-Feature Requirements for Voice over Data Applications (Continued)

Feature	Description	Requirement in ATM	Requirement in IP Net	Requirement in Frame Relay
Voice switching	Data network equipment can generally support on-net applications. Off-net is also critical. At the very least, the adjunct must decide whether to route a call over the internal data network or to route it to the public switched telephone network.	Ability to route off-net is a must for real applications.	Ability to route off-net is a must for real applications.	Ability to route off-net is a must for real applications.

tariffs and/or pricing arrangements that provide voice services for less than $0.08 per minute. Somewhere between the traditional and untraditional extremes one finds voice over frame relay; further toward the untraditional end is voice over ATM; all the way in the untraditional camp one finds voice over IP (for intranet/enterprise network applications) and voice over the Internet (for geographically dispersed applications). Bandwidth efficiency and quality are the principal trade-offs in this arena. Besides traditional telephony, voice over frame relay currently has the best chance to be used for commercial applications during the next couple of years.

Standards and products for voice over data networks are emerging because of a change in the economics of both private networks and public networks. Currently a number of organizations have significant investments in private data facilities that have capacity available to carry additional on-net traffic with what is perceived to be little initial incremental expense; on the other hand, many corporate networks are already very congested (globally or at least in some segments) for data transmission, and so planners would be ill advised to add additional flows.

Figure 11.2 depicts some of the key voice over data technologies available. The protocol stack approximately represents the kind of functionality to be available in the interworking/adaptation device. The figure shows

■ Voice over frame relay

■ Voice over frame relay service interworked to ATM

- Voice over ATM with AAL1/CBR (constant bit rate), AAL5/VBR (variable bit rate), or the newly introduced AAL2/VBR.

- Voice over IP carried in a frame relay network (e.g., LAN-originated, and over a WAN)

- Voice over IP in a LAN

- Voice over IP over the Internet

- Voice over IP carried in an ATM network [using one of three available IP-carrying methods: classical IP over ATM (CIOA), LANE, or MPOA].

There are two technical approaches for voice over ATM: one is via the constant bit rate service class and the use of AAL1, and the other is via the variable bit rate service class and the use of AAL5. The CBR approach is currently the more common of the two approaches, and has been around longer. One of the advantages is that it provides circuit emulation service (CES), meaning that it gives a PBX the appearance that a T1 line is available, while in reality there is an ATM

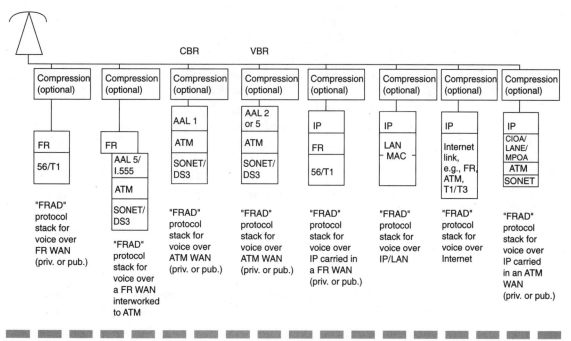

Figure 11.2
Voice over data networks. CIOA, classical IP over ATM; LANE, LAN emulation; MPOA, multiprotocol over ATM.

permanent virtual connection (PVC). The disadvantages of this approach are that (1) there is a need for higher bandwidth to recover a DS1 signal, (2) a mesh of end-to-end point-to-point emulated circuits is required, generally without (carrier) network participation, and (3) there is limited support of traditional voice telephony signaling; this in turn implies that voice is supportable only for on-net applications—namely, only the users at the various termination points of the organization's enterprise network are accessible; other voice users are not accessible.

On frame relay and/or IP networks, capabilities are needed to transform best-effort communications into functionality that can support both streaming voice and bursty data traffic. ATM networks may or may not be utilized for best-effort services. Originally ATM was designed to support CBR and VBR traffic types, with user-to-network traffic contracts. More recently, the computer industry has requested that best-effort services such as available bit rate (ABR) and unspecified bit rate (UBR) be added (currently only a few switches/carriers support ABR/UBR).

In general, the economics of voice over ATM are not yet favorable, unless an organization already has an ATM-based enterprise network that uses ATM via edge multiplexers, and the organization is interested only in on-net voice. Today, the typical access device ("edge multiplexer") is actually an ATM-ready router that supports legacy ports on the user side and ATM on the network side. Many of these devices do not support circuit emulation. However, some edge multiplexers now entering the market do support T1 interfaces on the user side, to connect PBXs utilizing circuit emulation.

As noted in Chap. 10, some carriers are looking more fundamentally to use ATM totally internal to their network as a statistical multiplexing technology to derive higher SONET efficiencies; however, significant rollout is not expected any time soon. This use of ATM (see Fig. 11.3) is transparent to the end user, but it would be the internal infrastructure used to carry voice. The bottom portion of the figure depicts a new kind of device now under development at some of the major network equipment providers.

11.2.2 ATM Technology Relevant to Voice

In the 1980s, organizations built large networks to accommodate specific kinds of transmission. These included voice networks, data networks, and video networks. This led to both a duplication of effort and suboptimal

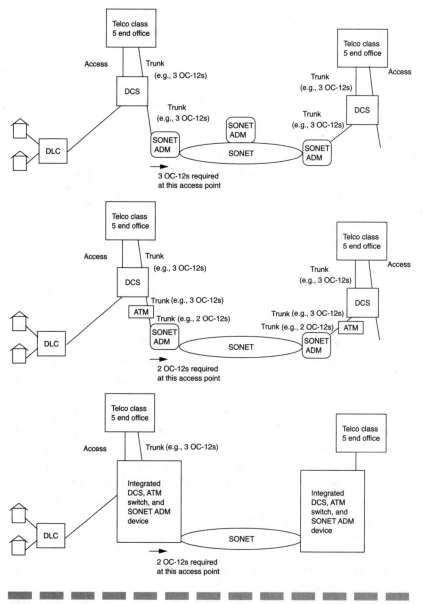

Figure 11.3
Use of ATM internal to network. *Top:* today's environment. *Middle:* evolving environment. *Bottom:* longer-term environment.

financial outlays. Because many of these networks were built for peak load conditions, the average utilization tends to be low, leading to financial suboptimality. Consequently, it became important for organizations to find ways of using a single network infrastructure and assigning bandwidth on an as-needed basis. ATM lets both private organizations and public service providers build "unchannelized" networks to make more efficient use of the underlying bandwidth on the network.

Currently, many organizations require transmission services above DS1 rates, but not yet as high as DS3. By offering scalable rates from 1.5 Mbps to 155 Mbps or higher under traffic contracts, ATM services can make the WAN transparent, in the sense that no speed reductions are required over the Internet (lowercase).

ATM is based on connections, not channels. The term *asynchronous* refers to the way in which ATM achieves its unchannelized bandwidth allocation. ATM sends information associated with a connection only when there is actual information to send. This is in contrast to channelized or time-division multiplexed (TDM) networks, where even if a channel is idle, a special bit pattern must be sent in every time slot representing a channel. The ATM cell is 53 bytes long, consisting of a 5-byte header containing an address and a fixed 48-byte information field or payload (frame relay uses a 2-byte header and a variable-length information field).

ATM functionality corresponds to the Physical Layer and a portion of the Data Link Layer of the Open Systems Interconnection (OSI) Reference Model. The ATM protocol functionally must be implemented in user equipment (e.g., hubs and routers) and network elements such as switches. The network does not participate in the end-to-end application that is running over an ATM connection after the connection has been established, since it operates only at the Data Link Layer.

The ATM cell size is a compromise between the long frames generated by data communications applications and the short, repetitive needs of voice traffic. Thus ATM allows a mix of data, voice, and video within the same connection or application.

ATM supports a number of service classes to handle the various information types on a network, as discussed in Chap. 2. This ensures optimal network usage and guaranteed end-to-end delivery. ATM naturally differs from TDM, which is currently the de facto method of carrying voice in the public switched telephone network. TDM is a synchronous mode of data communication that digital telephone networks have used since the early 1960s. These networks move information in fixed-length

8-bit "time slots." Because TDM networks are designed to carry voice, 8000 time slots per second (the rate of sampling used for the digitalization of voice) are allocated to each connection. This concept is what creates the basic 64-kbps single digital channel. Transferring information in time slots works only if the time slots are synchronized. Information arriving on a node can expect to find its allocated time slot waiting to carry the information out. If input and output time slots are skewed from each other, the output slot might "depart" before it was filled. Network clocks are synchronized to assure that time slots match throughout the network.

Unlike TDM, ATM attempts to allocate bandwidth more efficiently by giving users access to the entire communication channel on demand. If the channel is in use, a user may have to wait to gain access. However, as cells are small compared to frames and of fixed length, the delay is minimal and can be controlled. The technology is not really new. People were using statistical multiplexing in the late 1970s. Hence, one can say that ATM is statistical multiplexing where the entire industry has agreed (by developing standards) on how to do it.

All of the ATM layers discussed in Chap. 2 have be taken into consideration for voice, but particularly the ATM Adaptation Layer. As implied earlier, the debate is over whether to use AAL1 and circuit emulation service (that is, CBR), or evolving AAL2 and real-time (rt) VBR.

11.2.3 Voice Compression and Approaches

The quality of voice transmission is determined both by perception and by measurement (quantization noise, etc.). Voice transmissions can tolerate no more than 350 to 400 ms of delay round-trip at most; in fact, for traditional commercial applications, that delay has been of the order of 10 to 30 ms. For voice over data networks, occasional dropped packets, frames, or cells are not an issue, since the human ear can tolerate small glitches without losing intelligibility.

Current approaches for voice over ATM assume a PCM model in which voice encoding and transmission takes place in real time. In effect an entire DS1, composed of up to 24 voice channels, is transported end to end over ATM using *network interworking* techniques. This model imposes a need to preserve timing in frame delivery and playback; this can be accomplished with a kind of time stamping in the AAL1 header.

However, at this juncture, voice over data networks in general (and frame relay in particular) have taken a different approach: A compressed and packetized model of voice transmission that separates the time scales for encoding, transmission, and playback has emerged. Hence, preserving synchronous timing is no longer necessary; improvements in encoding algorithms and faster and cheaper hardware have changed the paradigm. At this juncture, most voice encoding uses some kind of prediction technique at the receiving end. Predictions are based on the most recently received information. Therefore, if a frame is lost, the newly arriving frame will show that the receiver's prediction is not current (since the receiver was not updated by the missing frame). It follows that the output is not correct, and the result is distorted voice. Hence, the performance is related to both delay and loss in the frame relay network. The issue is how much time is needed for the receiver to catch up with the arriving frames and get current, so that the voice output will be as intended. "State-of-the-art" voice compression algorithms of the early 1990s could require several seconds to synchronize after a loss of bits. Newer algorithms are able to self-synchronize within the length of a single frame. This makes each frame effectively independent: Since human ears can compensate for the loss of 20 (ms) of sound, an occasional lost frame does not disrupt communications (however, if every other frame were lost, then there would be a serious problem). There are many FRADs on the market that now support voice switching, fax demodulation, echo cancellation, silence suppression, and dynamic bit rate adaptation technologies with an exact implementation of ITU-T's new G.729 standard voice algorithm.

As related to ATM, there is not yet much application of compressed voice, but this could come in the next couple of years. In 1995 the ITU-T standardized the ACELP (algebraic code excited linear prediction) voice algorithms for the coding of speech signals in wide area networks. ACELP and code excited linear prediction (CELP) are used for compression rates below 16 kbps. ITU-T G.729 (CS-ACELP) is an international standard that compresses the standard 64-kbps PCM streams as used in typical voice transmission to as low as 8 kbps. ITU-T G.728 (called low-delay CELP or LD-CELP) is an international standard that compresses to 16 kbps. ITU-T G.723.1 compresses voice to rates as low as 5.3 kbps. Many FRAD vendors offer proprietary algorithms that drop the rate down as low as 2.4 kbps.

There have been some recent efforts aimed at standardization of voice over ATM. As of press time, CES-based services were standardized and AAL2 approaches were under development.

The ATM Forum started work on voice transport in 1993, and it was not until April 1995 that the VTOA (Voice and Telephony Services over ATM) Working Group published its first document, which contained the unstructured circuit emulation and structured circuit emulation specifications. Unstructured circuit emulation maps an entire T1 (1.544-Mbps) circuit to a single ATM virtual circuit, thus limiting it to point-to-point applications. On the other hand, structured circuit emulation allows switches to map individual 64-kbps circuits in a T1 line to ATM virtual channels (VCs), and it can be used for point-to-multipoint connections. Each requires voice to be treated as CBR traffic. A problem is that CBR traffic forces customers to reserve bandwidth for voice even when they're not actually sending it.

Sending voice as VBR traffic is the obvious alternative. There is work underway to develop standards for VBR support of voice. Silence suppression and voice compression will be a part of the new specification, providing greater use of bandwidth. AAL2 is a proposal in ITU-T Study Group 13. Initially AAL2 targeted VBR service for low bit rate voice traffic between a wireless base station and a mobile switch center, but it could be parlayed into a new voice standard for more general applications.

Work is focused in two areas: VTOA trunking for narrowband services, targeted primarily at applications in private voice networks, and VTOA legacy voice services at a native ATM terminal, targeted at applications in private and public networks, where interworking and interoperation of ATM and non-ATM networks and services for voice are necessary (see Table 11.2).

Circuit Emulation Service Interoperability Specification was released in 1997. This service has a CES and a dynamic bandwidth CES version. Features are as follows [em]:

CES:
- Constant bit rate (64-kbps) service
- Structured service for fractional applications (FT1/FE1)
- Unstructured service for DS1/E1/DS3
- Options for carrying channel associated signaling (CAS)
- Configured bandwidth used whether there is traffic or not
- More costly than TDM solutions because of ATM's overhead

TABLE 11.2	**Work Item**	**Document Number**	**Status**
ATMF VTOA Specifications	1. Circuit Emulation Service (CE-IS v2.0)	af-vtoa-0078.000	Approved January 1997
	■ Structures and unstructured service		
	■ Sync. and async. clock recovery		
	■ ATM CBR PVC (SPVC) service		
	2. Dynamic Bandwidth CES (DBCES)	af-vtoa-0085.000	July 1997
	■ Defines DB IWF based upon CES-IS		
	■ PVC and optional SVC services		
	■ Idle pattern detection		
	■ CAS AB bit detection		
	3. Landline Trunking over ATM	af-vtoa-0089.000	July 1997
	■ B-ISDN to N-ISDN (public) interworking		
	■ ISDN CCS support/ interpretation		
	■ $N\times64$-kbps voice switching (ATM SVCs)		
	4. ATM Desktop	af-vtoa-0083.000	Approved May 1997
	■ B-TE to B-ISDN interworking		
	■ B-ISDN to N-ISDN interworking (public and private)		
	■ AAL5 and AAL1 B-TE options		
	■ 64-kbps voice switching (ATM SVCs)		
	■ UNI 4.0 supplementary service		

Dynamic bandwidth CES:
- CES structured service models FT1/FE1
- Bandwidth allocated dynamically based on active channel indication
- CAS or common channel signaling (e.g., ISDN's D channel)
- User required to allocate bandwidth for the maximum number of channels

The *Trunking for Narrowband Services* specification is based on the use of an interworking function (IWF) between the ATM network and each interconnected narrowband network. This includes a Land Line I, a Land Line II, and a Mobile Trunking specification.

Land Line I

- Targets ISDN trunking over an ATM network
- Uses call-by-call routing to make effective use of bandwidth
- Supports dynamic bandwidth by allowing VCC setup in response to call
- Provides 64-kbps channels
- Has no silence suppression

Land Line II

- Targets access and private trunking applications
- Uses call-by-call routing to make effective use of bandwidth
- Uses new AAL2 to multiplex multiple channels in a cell
- Supports compressed voice and silence suppression
- Supports switched calls to outgoing trunks
- Supports fax

Mobile Trunking

- Targets cellular networks
- Supports transport between the base station and the mobile service central office
- Efficiently transports voice that is already compressed by mobile handset
- Uses AAL2 to multiplex multiple channels into a cell

The *Voice and Telephony over ATM to the Desktop* specification specifies the functions required to provide voice and telephony services over ATM to the desktop. It describes the functions of the IWF and a native ATM terminal. This version covers only the transport of a single 64-kbps A-law or mu-law encoded voiceband signal.

11.2.4   Advantages of the Integrated ATM Data/Voice Approach

In principle, the emerging technologies for transmitting voice over data networks present opportunities for organizations to reduce costs and enable new applications. In particular, traditional router vendors see the opportunity to cannibalize the voice traffic by adding features to enable their routers to provide any of the voice over data flavors shown in Fig. 11.2. Clearly, if a company uses separate facilities to carry on-net voice (company location to company location), there could be additional costs in terms of communication channels, equipment, and carrier charges.

Enterprises used to justify the cost of a private WAN by the cost savings these networks achieved for the on-net voice traffic. Now, bandwidth requirements for data networks are so great that organizations can add voice capabilities to these networks for relatively limited incremental costs. These considerations apply primarily to voice over frame relay; voice over IP will experience quality degradation, and voice over ATM is relatively expensive at this time, in terms of the required CPE.

Figure 11.4 depicts a voice over ATM application when the organization uses an ATM carrier that has an overlay ATM network (namely, the ATM network and the voice networks are separate). Only on-net voice

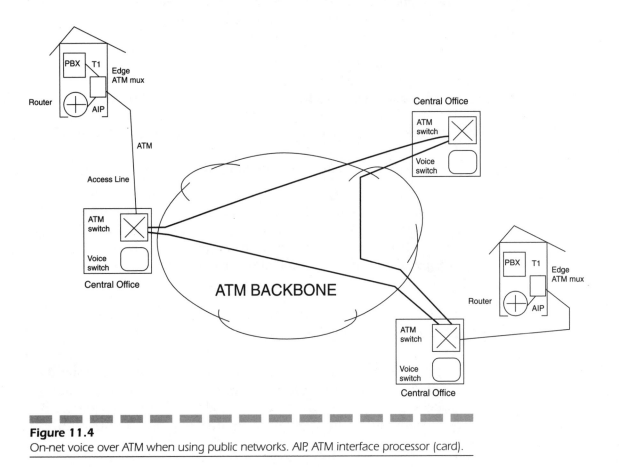

Figure 11.4
On-net voice over ATM when using public networks. AIP, ATM interface processor (card).

applications are supported. Note that the organization needs an ATM edge device that supports circuit emulation, so that the PBX can be plugged in. In some cases the organization already has an ATM network that may not support circuit emulation. This is typically the case when the organization utilizes ATM to interconnect traditional routers. This would then necessitate an upgrade. Fortunately, new routers, such as Xylan 3WX, support both routing function and circuit emulation. Another option would be for the PBX to develop a direct ATM/AAL5/VBR connection to the switch, but this may be more popular in a year or two.

There are no voice quality issues when using circuit emulation, particularly if the number of switched hops is small (say, less than three). This configuration makes sense if the organization has an ATM with (1) a CES-ready edge device and (2) available bandwidth in the network. Note that although the overlay arrangement limits voice access to on-net, one can hope that at some point the carrier would connect the two central office switches, achieving required connectivity.

Figure 11.5 shows essentially the same architecture, except that the ATM network is privately owned rather than being supplied by the carrier. The advantage of this approach is that the network planner can (perhaps) better control QoS and congestion, thereby achieving better quality. The disadvantage of this approach is that this will preclude any off-net voice access. Even if the carrier eventually offers connectivity to the public voice network, the user has no access.

Figure 11.6 depicts an integrated arrangement (currently not offered by any carrier) that allows on-net and off-net voice over ATM services. Although the arrangement in Fig. 11.6 is technically feasible, the economics and the commercial prospects are unknown. Note that at least in the user plane (the information-carrying plane), interconnection is achieved by putting the interworking function either in the ATM switch or in the voice switch. With the former, the voice switch remains unchanged; with the latter, the voice switch needs to be upgraded. Another alternative would be to upgrade the voice switch to support VBR/AAL5 capabilities, thereby eliminating the need for circuit emulation. The challenge in the design depicted in Fig. 11.6 is in the control plane, to achieve signaling-to-signaling conversion and support.

Voice and data traffic have different requirements for network services. A voice transmission requires only a small amount of bandwidth, but that bandwidth must be available on a dedicated (continuous) basis, with very little delay, delay variation, and loss. Even delays in the millisecond range can give rise to a noticeable echo or gap in the conversation. By contrast, data traffic can adjust to network delay and, with its bursty requirement,

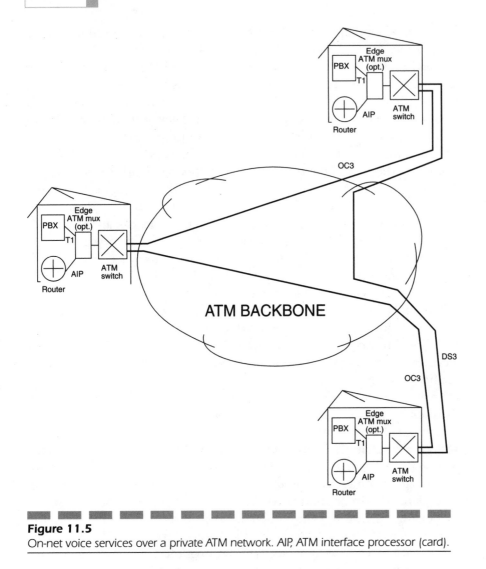

Figure 11.5
On-net voice services over a private ATM network. AIP, ATM interface processor (card).

can use the amount of bandwidth available in the network at any moment (e.g., TCP can throttle to match the actual throughput of the network).

Proponents offer the argument that in order for carriers to make ATM ubiquitous and inexpensive, it must support voice and multimedia. Also, if ATM switches are to replace T1/T3 multiplexers in private networks, they have to support voice. In fact, a number of ATM switch manufacturers, notably Nortel, Cisco/Stratacom, and GDC, are moving in that direction. Many switches now can carry AAL1 traffic with the desirable QoS features. One of the good things about this design, is that the voice quality is absolutely top-of-the-line. However, these switches carry AAL1 traffic only at the T1 level. Treating the entire T1 PBX-to-PBX tie line connection as a single class A VC implies that a separate link for voice for each pair of locations of interest is required. Furthermore, there may be limited support for actual PBX interfaces (beyond the basic bit stream and connector type). Also, this limits the voice to on-net applications, because there is no connection to the voice switch at the

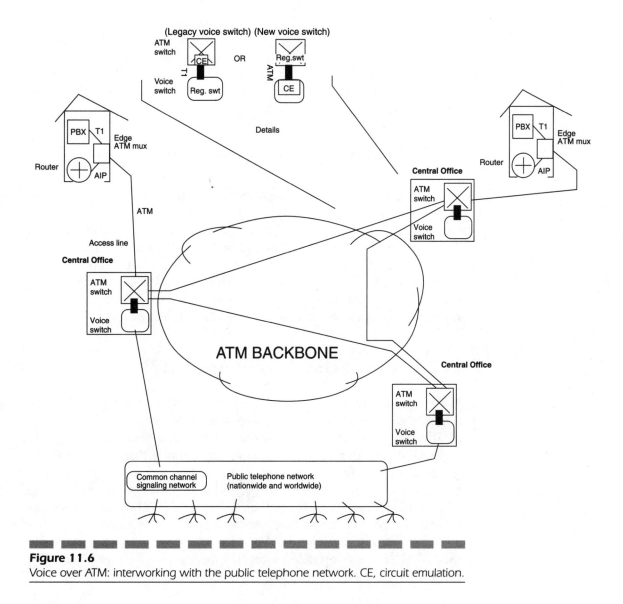

Figure 11.6
Voice over ATM: interworking with the public telephone network. CE, circuit emulation.

central office (as shown in Fig. 11.6). A first step in the direction of increased granularity is to develop the more sophisticated version of AAL1 support to be included in CPE: This is the *structured* AAL1, which allows individual DS0s to be identified (in the early *unstructured* AAL1, there was no DS0 visibility).

Proponents make the case that the support of voice via AAL1/circuit emulation is not really practical, and that AAL2/VBR/compression support is needed. Specifically, four voice processing enhancements are required to make voice over ATM attractive: voice compression, silence suppression, idle channel cell suppression, and signaling support, including translation of voice signaling to SVC ATM signaling. This will enable the use of AAL2 and VBR, which promises to be more bandwidth-efficient than CBR/circuit emulation methods. Some of these capabilities may appear in the ATM context in the next couple of years.

11.2.5   The Voice over ATM CPE Market

As discussed in the previous sections, CPE is required to support voice-to-ATM interworking. Namely, equipment is needed that accepts PBX trunks and, hopefully, legacy LAN links on the organization side, and ATM on the network side. This equipment will support either circuit emulation (CBR) via AAL1, VBR via AAL5, VBR via AAL2, and possibly compression. Such equipment is entering the market, particularly for CBR-based applications. Table 11.3 depicts some top-of-the-line CPE that supports routed services over ATM as well as circuit emulation.

11.2.6   Conclusion: Voice over ATM

The integration of several media is desirable because of the potential economic advantages. However, today voice over ATM remains more of a technical novelty. The likely prognosis is that over the next couple of years, more ATM services and technologies to support broadband internetworking of data networks will be introduced by organizations. As ATM becomes more widespread in the organization, and as CPE supporting voice-to-ATM interworking (whether CBR-based or VBR-based) becomes more prevalent and cost-effective, there is a possibility that voice over ATM will experience production-level and/or commercial deployment. Voice over ATM first was modeled after TDM-type solutions (i.e., CES). The advantages for carrying voice over a fastpacket network are in the VBR

TABLE 11.3 CPE That Supports Circuit Emulation

	Cisco (7204)	NetEdge (40)	NetEdge (85)	Sonoma Retix	Xylan	ODS (warrior1)	ODS (warrior)	3Com (2500)	3Com (2000)	CASCADE (SA-100)	CASCADE (SA-600)	LUCENT (AAC-3 FS)
Interfaces												
Ethernet	Yes	Yes	Yes	Yes	Yes	Yes	Yes	Yes	No	No	No	No
Fast Ethernet	Yes	Yes	Yes	Yes, 1 port	Yes	Yes	Yes	Yes	No	No	No	No
Token ring	Yes	No	Yes	Yes, 1 port	Yes	No	Yes	No	Yes	No	No	No
FDDI	Yes	No	Yes	No	Yes	Yes	Yes	Yes	Yes	No	No	No
ATM DS3	No	Yes	Yes	Yes, 1 port	Yes	Yes	Yes	No	No	No	No	Yes
ATM OC-3 SM (LR, IR)	Yes	Yes	Yes	Yes, 1 port	Yes	Yes	Yes	No	No	No	No	No
ATM OC-3 MM	Yes	Yes	Yes	Yes, 1 port	Yes	Yes	Yes	Yes	Yes	No	No	Yes
IMA	No	No	No	No	No	No	No	No	No	No	No	Yes
Circuit emulation	Yes	No	No	Yes	Yes	No	No	No	No	No	No	Yes
Standards Compliance												
IISP	No	No	No	No	No	No	No	No	No	No	No	No
P-NNI	No	No	No	No	Yes	Yes	Yes	No	No	No	No	No
UNI Version 3.0	Yes	Yes	Yes	Yes	Yes	Yes	Yes	Yes	Yes	No	No	Yes
UNI Version 3.1	Yes	No	No	Yes	Yes	Yes	Yes	No	No	No	No	Yes
UNI Version 4.0	Yes	Yes	Yes	Yes	Yes	?	?	?	?	No	No	No
IEEE 8021d	Yes	Yes	Yes	Yes	Yes	Yes	Yes	Yes	Yes	No	No	No
IEEE 8021i	Yes	Yes	Yes	Yes	No	No	No	Yes	No	No	No	No
IEEE 8023	Yes	Yes	Yes	Yes	Yes	Yes	Yes	Yes	No	No	No	No
IEEE 8025	Yes	Yes	Yes	Yes	Yes	Yes	Yes	No	Yes	No	No	No
RFC 1483 bridging	Yes	Yes	Yes	Yes	Yes	Yes	Yes	Yes	No	No	No	No
RFC 1577 IP over ATM	Yes	Yes	Yes	Yes	Yes	Yes	Yes	Yes	No	No	No	No

space. The greatest promise for effectively utilizing ATM is in the Land Line Trunking II and the Mobile Trunking ATM Forum specifications.

11.3 Voice over Frame Relay (over ATM)

Until a couple of years ago, planners did not see frame relay as supporting corporate voice requirements on a commercial basis. This is because frame relay technology was initially developed to support data applications, particularly router interconnections and the movement of system network architecture (SNA) traffic. Four key problems affected voice delivery over a frame relay network: (1) lack of standards for voice over frame relay, (2) lack of QoS support in frame relay, (3) poor voice quality and limited equipment availability, and (4) high overbooking of bandwidth on the part of carriers, further affecting the quality achievable over a WAN. However, improvements in all of these areas have been seen in the recent past, making voice over frame relay in the corporate enterprise network a practical possibility at this time. Voice over public frame relay networks is still at an early stage, although an increasing number of carriers are entering the business. On the plus side, one finds cost-effectiveness, particularly for integrated delivery of voice and data services to the SoHo segment. Voice support eliminates the need for multiple distinct facilities to and from remote locations. On the minus side, one finds continued concerns about the quality of the resulting voice.

Proponents saw 1997 as the "Year for Voice over Data." Equipment supporting voice, specifically voice-enabled FRADs, is now available from over a dozen vendors. Voice over frame relay is only one such approach; other approaches include voice over ATM, discussed earlier, and voice over IP networks. However, quality, ubiquity, interworking, and reach remain serious limitations for the foreseeable future for all these techniques. This section discusses the positioning of voice over frame relay in an enterprise network. It provides a brief synopsis of frame relay, a review of compression methods, and some vendor and market information.

11.3.1 Enterprise Networking Applications

By the mid-1980s, many companies realized that they (1) needed high-capacity digital connectivity to support companywide access to critical business information, and (2) needed to reach all company locations, including dispersed branch offices. This led to a transition away from dedicated point-to-point transmission facilities, such as T1/DS1 lines, and to the introduction of switched services such as frame relay.

With the increased interest in full-featured, companywide connectivity for enterprise networks and intranets, access technology for SoHo support is playing an ever more important role in the total communication solution that corporate planners are seeking to deploy at this time. Specifically, there is interest in integrated multimedia access technology: Users now look for equipment that supports data, LAN, voice (compressed), videoconferencing, and fax applications over public or private frame relay networks. Remote branch offices are prime candidates for such "entry-level," yet significantly important connectivity equipment. Given the fact that there may be many remote locations, the cost-effectiveness of the equipment and bandwidth efficiency are very critical.

A 1996 report commissioned by the Frame Relay Forum confirmed that frame relay has become the predominant method for the wide area interconnection of corporate enterprise networks. The growth rate in the number of worldwide users was 300 percent a year for the 1994–1996 period: The number of organizations using it grew from 1500 in 1994 to 15,000 in 1996. Also, the same study found that the most common port speed is still 56/64 kbps. Overall, the market is expected to grow at about 25 percent a year for the next 4 years. Hence, the support of voice over frame relay is a relevant design question. Frame relay has evolved from a solution for transmitting legacy and LAN data, to handling more delay-sensitive protocols (SNA), to the current challenge of delivering voice and fax across WANs to branch offices.

Figure 11.2 depicted some of the key voice over packet technologies available. The protocol stack approximately represents the kind of functionality to be available in the FRAD. Voice carriage over frame relay does not generally rely on the PCM techniques used to digitize voice since the early 1960s. Such techniques require 64,000 bps, based on Nyquist's theorem and dynamic range/quantization considerations. PCM uses a constant stream of bits in a TDM channel. The DS0/DS1/DS2/DS3 digital hierarchy still used in the United States is based on this technology. Frame relay, on the other hand, is a form of statistical frame-based multiplexing, with associated possibilities for frame-to-frame delay variation, frame loss, and jitter.

Voice over frame relay makes sense over an enterprise network (even if implemented with public carrier services), rather than for generic off-net applications. Furthermore, it makes the most economic sense for calls between a company's domestic and international sites.

Demand for integrating voice networks with data via frame relay started picking up speed in late 1996. The number of announcements from carriers in the recent past means that customers are pushing them for voice over frame relay services. Ameritech recently joined MCI and AT&T by announcing voice over frame relay service.

Frame relay services can be secured over private or public frame relay networks. The key specifications for the service are ANSI's T1.617 and T1.618. The Frame Relay Forum has also developed a set of critical implementation agreements (IAs) (see below). This successful industry consortium continues to promote the use of frame relay products and services around the world, building on the success of frame relay in North America and the growing markets in Europe and Asia. Initially these standards did not include intrinsic voice support; however, voice can be accommodated once a compression scheme is identified (so as to achieve greater packing density) and appropriate Quality of Service parameters for the frame relay service are established (since 1997 a voice over frame relay standard has become available). Many companies rely on public frame relay services from companies such as AT&T, Sprint, and Worldcom/MCI. New pricing trends are also giving impetus to the voice-support question: Frame relay service charges have recently decreased, after several years of stability. For example, in 1996, some carriers decreased costs by up to 10 percent. In addition, AT&T now includes frame relay in its Tariff 12. This is important because companies have been swapping out private line services in favor of frame relay; however, because of the more favorable economics of frame relay, and because frame relay revenue was not counted toward the company's revenue commitment, this in effect worked against Tariff 12.

Port-speed tariff prices for long-haul PVC service are approximately $275 for 56/64 kbps, $850 for 256 kbps, $1370 for 512 kbps, and $2300 for 1.544 Mbps. "Guaranteed" bandwidth [committed information rate (CIR)] is charged in multiples of 8, 16, 32, or 64 kbps. Some of the carriers are now also in the process of introducing SVC services and frame relay–to–ATM interworking.

As for the traditional approach to voice support, in the recent past long-distance carriers became involved in bidding contests that were driving down negotiated virtual private network (VPN) rates to as low as 5 cents per minute (as long as the organization was willing to channel all of its communication requirements to one provider). Reasons cited for this include anticipated concerns about the regional Bell operating companies' (RBOCs') entry into the long-haul market, and the emergence of voice telephony over packet networks (IP and frame relay).

The 5-cents-per-minute rate is available to customers with a $20 million a year minimum annual commitment (MAC); customers with a $1 million per year MAC can expect a 7.5-cents-per-minute rate. Some fastpacket vendors claim that organizations can achieve voice transmission at 5 cents per minute over frame relay, ATM, or the Internet. Users are reportedly leveraging this information when they deal with VPN carriers. Although carriers should not be concerned, planners are taking advantage of the perception.

11.3.2 Frame Relay Background

As noted, frame relay was originally developed to support the bursty traffic generated by LANs interconnected over WANs. In the early 1990s, it was used as a cost-effective consolidation of distinct corporate networks, such as legacy SNA networks. Frame relay's low cost compared to meshes of dedicated lines, along with the first traffic contracts available in the WAN market (ATM being the culmination in this respect), has made the service popular.

Frame relay is a statistical-supported frame-based connection-oriented service that provides transparent transfer of information on a best-effort basis. This implies that there is no guarantee of delivery; the network can discard frames in congestion situations, or when the user exceeds the traffic contract established with the carrier. For example, the user may be on a T1/DS1 line and contracts for 256 kbps of CIR with the carrier. But then the user could try to burst to a much higher level for a relatively long period of time. In general, users try to get away with the lowest possible CIR, even as low as 0, and then send more data. In that case, all of the relevant performance metrics, delay, jitter, and loss will be affected.

The frame relay PVC enables the delivery of frames across a WAN in the proper order, and error checking prevents delivery of frames with errors. However, frame relay service does not provide any intrinsic mechanisms for higher grades of service. In other words, there is no explicit method for establishing frame-level performance priorities. The only way to get any different traffic treatment is to

1. Set up different PVCs from the router/FRAD to the other end, over the WAN.

2. Administratively inform the carrier that certain PVCs carry "higher-priority traffic."

3. Hope that the carrier has a switch that has some level of different-grade-of-service treatment.

4. Hope that the carrier does not greatly overbook the bandwidth.

5. Hope that there are not too many hops in the network.

6. Acquire CPE (router or FRAD) that can itself handle different inputs (e.g., LAN segment, PBX, etc.) differently (according to some priority scheme).

Table 11.4 lists the key Frame Relay Forum specifications in support of the service, as of press time.

Although initially frame relay networks were separate overlay networks, today many carriers' frame relay networks are integrated with ATM. This means that the switch supports frame relay user-to-network interfaces (UNIs) to the user side, but internally it is an ATM switch. So, input frames are cellularized (typically on the input card), and the ATM Quality of Service and traffic/buffer management mechanisms are used to handle the derivative cells. On the output side, cells are either converted back to frames for immediate delivery or passed up to another switch for wide area coverage. At the destination switch, cells are converted back to frames

TABLE 11.4 FRF Specifications		
User-to-Network (UNI) Implementation Agreement		FRF.1.1; date: January 19, 1996
Frame Relay Network-to-Network (NNI) Implementation Agreement, Version 2.1		FRF.2.1; date: July 10, 1995
Multiprotocol Encapsulation Implementation Agreement (MEI)		FRF.3.1; date: June 22, 1995
Switched Virtual Circuit Implementation Agreement		FRF.4; date: N/A
Frame Relay/ATM Network Interworking Implementation Agreement		FRF.5; date: December 20, 1994
Frame Relay Service Customer Network Management Implementation Agreement (MIB)		FRF.6 (FRFTC93.111R3); date: March 1994
Frame Relay PVC Multicast Service and Protocol Description		FRF.7; date: October 21, 1994
Frame Relay/ATM PVC Service Interworking Implementation Agreement		FRF.8; date: April 14, 1995
Data Compression over Frame Relay Implementation Agreement		FRF.9; date: January 22, 1996
Frame Relay Network-to-Network SVC Implementation Agreement		FRF.10; date: September 10, 1996
Voice over Frame Relay Implementation Agreement		FRF.11; date: May 5, 1997

(having thereby achieved network interworking) or are delivered directly to the user over an ATM link. The user can in turn convert those cells back to frame relay (the advantage of this approach is that a high-speed link, e.g., 45 Mbps or 155 Mbps, can be used, rather than just a T1 link, as would be the case with a frame relay NNI). It should be noted that frame relay was not designed to support voice, whereas ATM has been developed from the beginning for this type of traffic mixing.

To support voice, it is necessary to support priorities in the switch (organization- or carried-owned). Hence, in looking at switch technology, one needs to ascertain how the priorities are allocated and managed. With top-of-the-line switches, such as Ascend/Cascade B-STDX 9000, Newbridge 36170, Cisco IGX-32 or BPX, Nortel Passport 160, ECI/Telematics NCX-1E6, and Sentient Ultimate 1000/2000, the frame relay priorities are supported via ATM priorities and Qualities of Service. Hence, a good understanding of the latter is needed. The bottom line, however, is that since ATM has strong Quality of Service capabilities, it has become possible to support priorities in frame relay, thereby making the support of voice over frame relay possible. Some switches support 2 frame relay priorities, others 4 or 8 priorities, still others as many as 32 priorities (these are assigned by administrative provisioning of the PVC). Table 11.5 depicts some of the desirable features of a state-of-the-art frame relay switch that enable it to support, among other services, voice. Some older and/or smaller switches do not support sophisticated priority mechanisms to handle PVCs in different ways (at the queue level).

TABLE 11.5

Key Features of Frame Relay Switch

Frame relay–to–ATM network interworking	Critical
Frame relay–to–ATM service interworking	Critical
Channelized DS3 plug-ins	Critical
Channelized DS1 for frame relay	Critical
ATM UNI DS3	Critical
CO power (—48 V)	Critical
Dual power source	Critical
Redundant switch processor	Critical
Frame relay point-to-point PVC support	Critical
Guaranteed CIR	Critical
Performance reports	Critical

TABLE 11.5

Key Features of
Frame Relay Switch
(Continued)

Billing reports	Critical
Telemetry access	Critical
Traps to multiple Network Management Systems (NMSs)	Critical
Local configuration database	Critical
Modularity	Critical
High port density	Critical
High backplane capacity	Critical
NEBS compliance	Critical
Year 2000 compliance	Critical
Load sharing	Critical
Dual power feeds	Critical
Hot-swappable (all elements)	Critical
1-for-N line redundancy	Critical
Support of standards (those in Table 11.3 plus others)	Critical
Large number of PVCs per port	Important
Frame relay point-to-multipoint	Important
Unchannelized DS1 plug-ins	Important
Configurable peak rate	Important
Oversubscription	Important
ATM traffic parameter support (PCR, SCR, MBS)	Important
ATM QoS classes	Important
NNI connectivity for frame relay	Important
External timing	Important
NMS support under HPOV	Important
Configuration GUIs	Important

Although frame relay is popular for its flexible bandwidth, its popularity has given rise to congestion in some networks, because carriers may not properly engineer their networks (in order to maximize their profitability) or have had difficulty keeping up with the demands for new service. Given just the demand for higher bandwidth as additional organizations subscribe, it is a relatively simple step for the public carriers to add switches and capacity to the backbone to accommodate the growing demand. But now customers are requesting lower-latency service for their carriers, so that voice and fax can be carried over frame relay. There is a perception that many carriers are struggling in their effort to increase capacity to meet this demand and/or to replace older switches with new ones that are ATM-based and support Quality of Service at the PVC level.

Carriers, central office switch manufacturers, and CPE vendors are now attempting to address the voice over frame relay opportunity simultaneously. Standards for sending voice traffic over frame relay are also emerging. CPE manufacturers have recently seen a growing interest in voice over frame relay, and have followed up with fast-paced research and the development of voice-capable FRADs. However, although the equipment choices are numerous, many of the carriers do not provide adequate support; only two companies, for example, now provide service-level guarantees for voice over frame relay.

11.3.3 Voice Compression and Approaches for Frame Relay

Frame relay has been positioned by proponents as the answer for organizations searching for an integrated service offering that will enable them to consolidate data, voice, and fax services into a single *WAN-access* mechanism (thus eliminating the access charges that have to be paid to regional Bell operating companies for traditional voice services between remote offices), as well as into a single WAN.

The conversion from analog to digital transmission in the early 1960s began an era of enhanced telephone communication quality and significant improvements in network utilization. As demand for telephone lines increased, even more efficient methods of using the existing transmission facilities were needed. The original standard algorithm for 64-kbps digital transmission was called pulse code modulation. Adaptive differential pulse code modulation (ADPCM) eventually improved efficiency to 32 kbps for each voice call, allowing twice as much traffic to be transmitted. PCM and ADPCM are currently used by telephone carriers worldwide to transport circuit-switched voice traffic [act].

The quality of voice transmission is determined both by perception and by measurement (e.g., quantization noise, etc.). As stated in Sec. 11.2.3, voice transmissions can tolerate no more than 250 to 300 ms of delay round-trip at most; in fact, for traditional commercial applications, that delay has been of the order of 10 to 30 ms. Voice delays in frame relay

TABLE 11.6

Various Voice
Delays

Traditional telephony networks	20–30 ms
Satellite networks	250 ms
Frame relay	125–193 ms
	Input buffer: 24 ms
	Compression: 20 ms
	Access queues: 0–24 ms (depending on network complexity)
	Network latency: 5–25 ms (depending on network complexity)
	Far-end queue: 0–24 ms (depending on network complexity)
	Jitter buffer: 72 ms
	Voice decoder: 4 ms

networks, when using voice compression, can be around 125 to 200 ms, as shown in Table 11.6. For voice over data networks, occasional dropped packets, frames, or cells are not an issue, since the human ear can tolerate small glitches without losing intelligibility (the retransmission of dropped packets containing voice information often is self-defeating). Although perceptions of voice quality vary by individual, the mean opinion score (MOS) is a widely accepted measure of voice quality. MOS ratings provide a subjective quality score averaged over a large number of speakers, utterances, and listeners. The following list indicates MOS values: 4.0 to 5.0, toll quality; 3.0 to 4.0, communication quality; <3.0, synthetic quality [act]. Digitization/compression algorithms are usually ranked based on this method, as noted later on.

Early approaches for voice over frame relay service assumed a PCM model in which voice encoding and transmission takes place in real time. This model imposed a need to preserve timing in frame delivery and playback; this can be accomplished with time stamping. Time stamping involves having each frame depict when it was originated; this enables the receiver to play back the frames with the same relative timing. This works in conjunction with a per-connection priority mechanism in the CPE/switch, so that voice frames receive higher priority and more predictable end-to-end delay variation. Several mechanisms for time-stamping voice frames have been proposed to the Frame Relay Forum and the ITU. The ITU has adopted a procedure for sending ADPCM over packet networks. The goal of these efforts was to maintain the locked-timing environment that evolved naturally from PCM transmission over a synchronous channel. However, at this juncture, voice over frame relay has take a different approach; a compressed and packetized model of voice transmission has emerged that separates the time scales for encoding, transmission, and playback, as discussed in Sec. 11.2.3.

The Frame Relay Forum's Technical Committee worked (during 1995–1997) on a specification that combines a number of different alternatives (see Table 11.7 for recent FRF activities, including voice over frame relay). The assumptions underlying the proposal are as follows:

1. Voice over frame relay will be compressed 8:1 compared to 64-kbps PCM, and most access/transmission lines operate much faster than the voice encoding rate of 8 or 16 kbps (e.g., at 8 kbps, a 40-ms interval of voice-encoded speech is 320 bits, which can be transmitted across a DS1 link in about 0.2 ms).

2. Voice will be packetized. While PCM does not allow or need the storing of voice, frame relay buffers voice segments at intermediate switching points.

3. The receiver will buffer frames in sequence and need not play them back immediately.

4. To the extent possible, voice must be carried over existing frame relay networks, particularly without modification to the existing standards.

5. Silence suppression is desirable.

6. Integral echo cancellation and delay compensation will preserve voice quality (low compression/decompression delay helps prevent "voice collisions" between speakers who both start talking at the same time).

During the early 1990s, significant advances in the design of digital signal processors (DSPs) occurred. A DSP is a microprocessor that is designed specifically to process digitized signals such as those found in voice and video applications. DSP development has allowed manufacturers to bring to market high-quality digitization algorithms that consume very little bandwidth. Voice compression algorithms make it possible to provide high-quality audio while making efficient use of bandwidth. The most commonly used voice compression algorithms are as follows [act]:

■ Pulse code modulation/adaptive differential pulse code modulation

■ Code excited linear predication/algebraic code excited linear predication

TABLE 11.7

Recent Frame Relay
Forum Directions

The membership of the Frame Relay Forum (FRF) had several goals at press time:

■ The Technical Committee and the membership (through ballot) to finalize the first phase of the *Voice over Frame Relay Implementation Agreement* (IA), and continue to enhance this exciting application beyond the 1997 accomplishments. The North American Market Development and Education Committee (MD&E) recently released a white paper on voice over frame relay (available at http://www.frforum.com).

■ The European MD&E Committee to continue to educate potential customers and providers about the benefits of implementing frame relay in Europe (e.g., the recent Frame Relay 2000 initiative in Europe, sponsored by the FRF).

■ The Board of Directors to begin the FRF's efforts to promote frame relay in the Pacific Rim, Asia, and China.

■ The MD&E Committees to work to bring frame relay into the mainstream of SNA business applications, through carrier services and advanced functionality in frame relay access devices and frame relay switches.

FRF.10, Switched Virtual Circuits at the Network-to-Network Interface, was completed in late 1996. In 1997, FRF members participating in the Technical Committee and associated working groups were focusing on:

■ FRF.11: Voice over Frame Relay (VoFR). This specifies procedures for the transport of packetized low bit rate voice over frame relay networks. This first phase, IA, defines support for preconfigured connections allowing multivendor interoperability. The VoFR IA became available in mid-1997.

New efforts initiated by the membership that may lead to IAs in the future include:

■ FRF.12: Frame Relay Fragmentation. This enables fragmentation and reassembly of frames at the FR UNI. This IA became available at the end of 1997.

■ Multilink Frame Relay at the User-Network Interface: This will enable frame relay devices to use multiple physical links; in essence, frame relay inverse muxing.

■ Frame relay–to–ATM (FR/ATM) Switched Virtual Connection Service Interworking: This effort will define FR/ATM service interworking supporting SVCs, popular in ATM, and already defined for frame relay.

■ Frame Relay Network Service-Level Definitions: This effort will provide FR network performance definitions. This will enable service providers to deliver service levels that are uniform across diverse networks. It will provide users with a metric for use when determining if contracted service levels were rendered.

■ Proprietary schemes, e.g., adaptive transform coding/improved multiband excitation (ATC/IMBE)

PCM and ADPCM, the traditional PSTN/PTT algorithms, receive high (toll-quality) mean opinion scores. MOS of 4.4 for PCM and 4.1 for

ADPCM are achieved by consuming 64 kbps bandwidth and 32 kbps bandwidth, respectively [act].

The ATC algorithm is actually a combination of time-domain harmonic scaling (TDHS), linear predictive coding (LPC), and vector quantization (VQ). The key features of the ATC algorithm are (1) low complexity and (2) a variable digitization rate. ATC has an MOS score of 2 to 3.8 depending on the voice digitization rate. The IMBE algorithm is also a hybrid coder. The underlying theory of IMBE coding is that various frequency bands in the speech spectrum behave differently with respect to a voiced/unvoiced classification. ATC consumes 8 to 16 kbps bandwidth, and IMBE consumes 2.4 to 8 kbps bandwidth, with MOS scores of "communication quality" [act].

ACELP grew out of years of study at various research institutions using CELP and CELP-like coders. The three main elements of ACELP are (1) linear predictive coding (LPC) modeling of the vocal track, (2) sophisticated pitch extraction and coding, and (3) innovative excitation modeling and coding. Independent tests indicate that the perceived quality of voice is equal to or better than the industry-standard 32-kbps ADPCM (G.721). ACELP is rated with an MOS of approximately 4.2. The recent introduction of ACELP allows "toll-quality" voice transmissions over frame relay networks. A variation of the ACELP algorithm is currently being reviewed by the ITU for Recommendation G.729 at 8 kbps [act].

Normally, complexity goes up and quality goes down as compression increases. The recent developments in very low bit rate voice digitization (e.g., ACELP), however, demonstrate that voice can be compressed as low as 4.8 kbps and still achieve near toll quality. With low-cost toll-quality voice compression algorithms and management of voice and data transmission parameters, voice quality can be maintained in high-traffic networks. Integrated FRADs need to employ techniques such as predictive congestion management, jitter buffers, fragmentation, variable rates, prioritization, silence detection, and digital speech interpolation, among others, to retain acceptable voice quality levels.

There are many integrated FRADs on the market. Starting in the early 1990s, a number of FRAD manufacturers supported voice compression, echo cancellation, silence suppression, and dynamic bit rate adaptation; fax and compressed data (4:1) were also supported. By the mid-1990s, many FRADs supported implementation of the ITU's G.729 standard voice algorithm (CS-ACELP) for compression rates below 16 kbps; however, interworking continued to be an issue. Also, many vendors used proprietary variations, and/or compressed voice to as low as 2.4 kbps. Some FRADs offered a choice of compression techniques. With the new FRF.11 specification, the issue of interworking should resolve itself in the 1998–1999 timeframe.

To ensure reliable voice packet delivery, the integrated voice/data FRAD must be designed to minimize congestion. Fragmenting data packets allows voice packets to traverse the network within acceptable delay parameters. Configuring network buffers to reasonably small depths permits rapid transmission of voice across many network-level queues [act]. Many vendors offer FRADs with advanced techniques to overcome delay problems. These techniques include:

Voice compression.   As already noted, ITU G.729 (CS-ACELP) is an international standard that compresses the standard 64-kbps PCM streams used in typical voice transmission to as low as 8 kbps. ITU G.728 (called LD-CELP) is an international standard that compresses to 16 kbps. Many vendors offer proprietary algorithms that drop the rate down as low as 2.4 kbps.

Echo cancellation.   ITU G.165 defines a method of dealing with echo, which occurs when delays cause voice traffic to be reflected back to the transmission point. Echo becomes perceptible when the delay exceeds 15 to 20 ms.

Traffic management.   Traffic prioritization algorithms in the CPE can place voice frames at a higher priority than data frames and can tune the size of the data frames to reduce delays for voice traffic.

Continuity algorithms.   These are designed to intelligently fill the void of missing or erroneous voice frames.

To produce a viable voice offering, carriers must eliminate significant delays, jitter, and loss through the backbone. Only the carriers that have replaced older frame relay switches and have carefully planned their network capacity will find this task possible. Upgrading older switches to higher-end second-generation equipment is not a trivial undertaking, but doing so allows the carrier to support the prioritization techniques (traffic-based PVCs), advanced bandwidth management, and congestion control needed to handle the load. The CIR provided by carriers may be of little consequence, particularly if it is low and/or if the carrier overbooks bandwidth excessively, because it is an average, and actual throughput can vary significantly from moment to moment. At press time, less than 1 percent of most service providers' frame relay traffic is voice.

TABLE 11.8

FRF Press Release on FRF. 11 (http://www. frforum.com)

The Frame Relay Forum today announced in May 1997 the ratification of FRF.11, an Implementation Agreement (IA) providing for Voice over Frame Relay (VoFR) communications.

Specifically, FRF.11—entitled "Voice over Frame Relay"—provides for bandwidth efficient networking of Voice and Group 3 fax communications over Frame Relay, as well as defining multiplexed Virtual Connections (VCs). This latter function allows for up to 255 voice and data sub-channels to be carried over a single VC through a Frame Relay network. Transparent relay of Group 3 fax communications is provided for in the IA by "spoofing" of the fax protocol, and transmission of fax traffic as a low-bit rate digital stream.

Two classes of voice compliance are supported, for maximum flexibility and worldwide applicability. Class 1 compliance calls for use of G.727 EADPCM typically at 32 Kbps (2:1 compression). Class 2 compliance specifies G.729/G.729A CS-ACELP at 8 Kbps (8:1 compression). In order to maximize use of bandwidth, it is possible to carry multiple voice samples in a single frame, further minimizing overhead.

The new IA is significant in that it is the first step for true multi-vendor interoperability of voice and fax over Frame Relay. Already offering a streamlined, compatible method for interoperability of data applications over the wide area, the new IA bolsters Frame Relay's ability to support the full complement of corporate LAN and legacy data, voice, fax and packetized video communications.

In May 1997, the Frame Relay Forum ratified a standard called FRF.11 for interoperable voice over frame relay networks (see Table 11.8). Up to now, users had to rely on vendor-specific solutions. Implementation plans are unclear at this time: Frame relay service providers have cautioned companies against implementing voice over frame relay, most ostensibly because these carriers also stand to lose voice traffic revenues. Secondarily it may be that their networks are unable to support the QoS required for voice. However, it is expected that equipment suppliers will implement the standard by early 1998.

11.3.4 Other Methods for Voice over Data Network

The Voice over IP (VoIP) group is a consortium of vendors backed by Intel and Microsoft set up to recommend standards for telephony and audio conferencing over the Internet. Its goal is to ensure interworking among vendors of Web-based telephony. At a 1997 meeting, the VoIP agreed to recommend the ITU-T G.723.1 specification. An alternative is ITU-T G.729A. Web telephones that can be used in conjunction with Internet services bypass the public telephone network. In addition to their poor quality, they suffer from the fact that they are proprietary. Hence the need for standards.

G.723.1 operates at 6.3 kbps. Key supporters are Microsoft, Intel, and the major videoconferencing vendors. One drawback is that it delivers limited voice quality. This standard is used in the H.323 standard for conferencing over LANs. G.729A operates at 7.9 kbps, and therefore is slightly better than G.723.1. Supporters include Netspeak, AT&T, Lucent, NTT, and France Telecom. G.723.1 is considered a good first step, and is best suited for intranets and controlled point-to-point IP-based connections. RSVP and available network bandwidth are required to ameliorate quality. New applications are now emerging. For example, some are adding Web telephone access to their call centers, letting customers reach the carrier's customer service agent by clicking an icon at their Web site that reads "speak to the agent." But in order to provide this on a broad scale, standards are required. RSVP features include

- Soft state, which enables monitoring of router information through periodic messages
- Receiver-controlled reservation requests
- Flexible control over sharing of reservations and forwarding of subflows
- IP multicast for data distribution

A related area deals with "streaming audio" over an enterprise network. Proponents say, "No sooner will your intranet be up and running than it will dawn on the human resources director, the chief financial officer, the CEO, or all three that they can send their monthly messages to the troops via audio or video." Streaming is a technique of breaking up a file into pieces and sending those to the user in a sequence. User software can begin to process the pieces as soon as it receives them. For example, a streaming system would break compressed audio data into many packets, sized appropriately for the bandwidth available between the client and the server. When the client has received enough packets, the user software can be playing one packet, decompressing another, and receiving a third. The Internet provided the impetus for the development of streaming technologies.

Streaming audio and video systems use an encoding process for compressing and packetizing the datastream and a decoding process for managing buffers according to available bandwidth, decompressing the packets, and rendering their contents. Most products available today are based on proprietary methods, limiting interoperability. Because coder/decoder systems usually aim at reducing the contents' data rate, they are "lossy." This implies that the quality of the contents will be degraded, to various degrees. Lossy compression is one of the reasons these systems are proprietary. Some examples of audio streaming products are Internet Wave (VocalTech, vocaltec.com), RealAudio (Progressive Networks, realaudio.com), StreamWorks (Xing Technology, xingtech.com), ToolVox (Voxware, voxware.com), and TrueSpeech Internet (DSP Group, dspg.com).

11.3.5 Advantages of Integrated Frame Relay (over ATM) Approach

Frame relay services are available almost anywhere, and companies could save money by consolidating voice and data traffic over a single network, particularly on the access side. Frame relay is also becoming available in the Asian and European markets. The voice traffic can be compressed to 32, 16, or 8 kbps using ITU standards. FRAD vendors also use proprietary algorithms that run at a number of speeds, including 4.8 kbps, 7.47 kbps, and 9.6 kbps. FRADs can cost in the $1200 to $5000 range, with a few costing less and a few costing more.

More generally, market figures show that voice/data muxes are "hot" now; worldwide sales were estimated at $410 million in 1995, and were expected to grow at about 20 percent a year through 1997 (much of that growth is expected outside the United States, as corporations extend their global networks to regions in which leased lines are expensive and/or in short supply).

In spite of the user interest and of the availability of CPE, most frame relay carriers are not prepared today for voice over frame relay and will not guarantee any level of voice over frame relay. Some carriers have no plans to offer this service-level guarantee ever. In fact, there are only two that offer it today, Intermedia Communications International (ICI) and Infonet Services Corp., and they have taken different approaches to addressing the performance problems.

The approach of ICI is to build its network with high-end switches, which allows prioritization, control and management functionality, redundancy, and fallback into the network backbone to protect the critical voice operations between branch offices. This approach is initially expensive, but it will let ICI offer CPE equipment that supports an upcoming voice over frame relay standard under development by the Frame Relay Forum. Utilizing second-generation equipment is not immediately practical for more established and much larger companies, such as CompuServe and the RBOCs, considering the amount of upgrading required to make a significant difference in their network's performance.

Another approach is to concentrate on the development of more efficient, but proprietary, methods to handle voice from within the CPE through the backbone. Infonet's Integrated Media Services (IMS), whose network service spreads to more than 30 countries, is taking this approach. It envisions better performance through its proprietary solution, Multi-Media Cell Technology (MMCT), developed in concert with Northern Telecom, which will include the technology in its future products. The use of proprietary solutions seems unsettling at first, but for some organizations the advantages of consolidated interbranch voice, data, and fax outweigh openness and CPE multisourcing.

What is also of interest is the method for billing for voice when the carrier explicitly offers such a service (the organization can always utilize its existing frame relay network and send voice over it, although the quality will be suspect). ICI offers a base rate of $300 per month, plus $10 for a common 56-kbps port, with a fee of $15 to $25 per month per location for management services. At the practical level, price breaks occur according to length of term agreements, revenue volumes, and other valued customer differentials. Infonet handles customers on an individual case basis, and makes a price quote after considering the traffic involved and any other specifics. Infonet charges for voice transmission on a per-minute basis, but targets a 25 to 45 percent savings over traditional voice carriers' prices, which it calculates and reports through its management service offering.

11.3.6 The Voice over Frame Relay CPE Market

Several CPE vendors are making a strong move into the multimedia (combined voice/data/low-end videoconferencing) access market. These vendors wish to introduce equipment that supports data, LAN, voice (compressed), and fax applications over public or private frame relay networks. There is a growing SoHo market for this kind of technology. Well-recognized vendors (such as Micom and Motorola), specialty developers (e.g., ACT Network and Memotec), and even large powerhouses like Cisco are vying for market share. A number of vendors, however, are approaching this market with low-end routers, rather than with traditional TDM-based multiplexers; routers tend to offer more network-level functionality in terms of synergistic enterprisewide internetworking cohesion and philosophy, particularly given the fact that the user's interface, even at the remote level, now is generally LAN-based. This approach to voice over data networks is worth watching and tracking, since it appears to be the clearly evolving approach to branch communication. However, routers do not handle voice and videoconferencing requirements well. Voice over IP is just evolving. Therefore hybrid devices that have both TDM and routing capabilities may be the best solution. Another version of this "hybrid" design is to include a capability to process data via the IP stack (and then via frame relay), and to handle voice only over the frame relay stack. (Refer again to Fig. 11.2.)

The 1995 worldwide FRAD revenues ($200 million) were as follows: Motorola ISG, 18.7 percent; Cisco, 14.5 percent; Ascom Timeplex, 14.3 percent; ACT Networks, 12.8 percent; Sync Research, 7.8 percent; Hypercom 6.2 percent; others, 25.8 percent. For voice over frame relay, the following are some of the key providers: ACT Networks, Cisco, FastComm, Hypercom, Micom, and Motorola. Vendors such as Motorola Network Systems Division and Hypercom Network Systems have introduced voice over frame relay products in the recent past. For example, Motorola's approach is to add voice support to its low-end Vanguard 100 and Vanguard 300 FRADs; a daughterboard and software are used with the Vanguard devices to handle voice on two ports at 8 or 16 kbps. This package is available for less than $1000. Hypercom Inc. built voice into its Integrated Enterprise Network (IEN) family of switch/routers, enabling users to integrate voice, legacy SNA, and LAN traffic over a single frame relay access line. Other vendors include Nuera Communications (San Diego, California), with the Access Plus F200.

In the TDM category (e.g., ISDN uplinks rather than frame relay), one finds, among others, RNS (Santa Barbara, California), with the NetHopper; Shiva (Bedford, Massachusetts), with the LANRover/Plus; SBE (San Ramon, California) with the SBE RouteMan; and Cubix (Carson City, Nevada) with the WorldDesk Commuter.

11.3.7 Access Technology: Generic FRAD Approaches

From a CPE-development perspective, available access technology, which also supports voice over frame relay, follows three general approaches:

1. "Low-end" multiplexers, with or without packet assembler/dissembler (PAD)-like functions. This approach is a continuation of the mid-to late

1980s/early 1990s trend of utilizing TDM methods, which originally had no higher-layer protocol support. Initially, CPE multiplexers provided a number of time slots that could be allocated to various subusers/ functions (e.g., 64 kbps or multiples thereof) on a TDM basis (with or without actual concentration on the user side). On the network side, this technology typically supported an aggregated T1/DS1 line—with no Data Link Layer protocol capabilities—that served these and other communication requirements. More recently, there has been the addition of frame relay Data Link Layer support on the network side; in effect, frame relay supports statistical multiplexing, making this newer equipment more bandwidth-efficient. In this case, the equipment provides frame relay PAD functions (i.e., FRADs). TDM-only architectures have the following limitations:

- Static bandwidth allocation
- Not bandwidth-efficient
- No switching capabilities
- Proprietary approach
- Not integrated with voice
- Low-bandwidth and low-throughput PADs and FRADs
- Developed primarily for data only
- No multimedia

2. Low-end routers. Many vendors are now approaching this market with low-end routers, rather than with TDM-based multiplexers. Routers tend to offer more network-level functionality in terms of synergistic enterprisewide internetworking cohesion and philosophy, particularly given the fact that the user's interface now is generally LAN-based, even at remote sites. Also, they offer an integrated-technology solution to the connectivity needs. Hence, it is more convenient to simply connect directly to the user's LAN—this implies a bridge and/or routing function. However, routers do not yet handle voice and videoconferencing requirements well (or at all), as noted, because of the limited voice over IP support at this time. There are efforts underway by router vendors to rectify this situation by developing effective queuing mechanisms. Bridge/router-only architectures have the following limitations:

- LAN is primary focus
- Inefficient handling of legacy traffic
- No multimedia support
- Fixed-configuration branch products are cheap but inflexible

3. Hybrid: includes both TDM and router (packet) technologies. This approach is employed (e.g., by Hypercom) when voice is an important component to be supported and/or when there is a need to support native SNA traffic (some corporate SNA traffic has already migrated to a frame relay format and/or an encapsulated IP stream). Hybrid systems combine both LAN and WAN routing into a single solution, so that all traffic is handled in its native mode. This allows all traffic to be separately prioritized. Bandwidth is not consumed by adding unnecessary overhead to encapsulated packets. Both of these factors are problems the pure LAN router vendors face. These considerations tend to be unique to branches and are differentiated from requirements of the core network. Hybrid products are designed to support the requirements of multibranch networks, which are significantly different from those of campus environments. These differences are

- Ongoing support for traditional serial applications
- Simultaneous support for LANs as applications gradually make the transition to client/server
- Lower traffic volume than the core of the network
- Integrated bandwidth generally limited to 56 kbps until voice/ video applications can be included and efficiently carried over the same transmission path

Hybrid systems can support LAN and serial simultaneously during the gradual migration to LAN client-server applications, optimize the flow of each of the different traffic types, and are price and functionality optimized for the volume of traffic at the SoHo. They do the job better than a simple router.

Another implementation of the "hybrid" is to support data over an IP/frame relay stack, and voice directly over the frame relay stack.

Planners must realize that there are these three competitive approaches at play. In general, routers now compete with TDM-based access devices for data-only environments. More traditional suppliers (e.g., ACT Network) follow the earlier and less sophisticated method. Some suppliers started out with multiplexers/FRADs and now have both products (e.g., Micom/NetRunner, Hypercom/IEN, Motorola/6520 MPRouter). Other suppliers (e.g., Cisco) have elected to take only the high-end router-based approach. *Note:* Some vendors offer bridging in conjunction with or in lieu of routing on the access device; in any event, a LAN-based packet technology is used in place of simple TDM and/or frame relay approaches.

The trend for the future appears evident: With the continued proliferation of LANs, the router-based enterprisewide/intranet approach is on the increase, and the multiplexer-based approach appears to be on the decline. There are some inefficiencies in encapsulating SNA into IP (e.g., with RFC 1490). However, many chief networking officers (CNOs) view standards support as critical, in order to be able to grow the corporation and the network (both in terms of size and in terms of new applications) in a straightforward manner. Some overhead is acceptable in exchange for flexibility; this is brought out by the fact that users are starting to utilize ATM, with its huge bandwidth inefficiency (12 to 15 percent).

Hybrid products are designed for high bandwidth efficiency when routing serial and LAN traffic over the WAN, by requiring less header overhead than routers. Hybrid products could outperform LAN routers if the user has the objective of minimizing tariff costs to the branch. These FRADs enable the migration from WAN-only to LAN/WAN applications with the same box, without creating parallel networks. Memotec should emphasize the advantages of hybrid technology as well as integration in its sales activities. Routing transports data across the network. LAN data require IP datagram routing, and WAN data require call-based routing. A sophisticated product supports all of these functions.

Eventually the router-based approach is expected to prevail. Router vendors are now in the process of bringing out a new series of equipment supporting voice and data for SoHo applications.

Table 11.9 depicts what the planner can expect in these kinds of products.

11.3.8 Conclusion: Voice over Frame Relay

The integration of several media at the branch/SoHo level is desirable because of the potential economic advantages for companies with many dispersed locations. However, today sometimes voice works well over public frame relay and sometimes it does not. It is important that network managers appreciate the potential problems associated with voice over frame relay, and how these problems can affect business transactions and caller satisfaction. In private frame relay networks, these issues can be addressed directly by properly upgrading to a second-generation switch and appropriately performing capacity planning. However, most organizations now rely on carriers to provide frame relay services. These carriers and the services they provide should be examined on a case-by-case basis to determine what kind of Quality of Service for voice traffic can be achieved over the network.

TABLE 11.9 Typical Integrated FRAD Features

	Momotec Products	Cisco Products	Hypercom Products	Motorola Products	ACT Products	Fastcomm Products	Micom Products
Product examples	CX900	1000s; 2500s (others in text)	IEN 500, 1000, 3000, 5000	Vanguard 100/200/300, 6500 Concentrators, MPRouter 6520, 6560	SMD-SF, SDM-JFP, SN 7040/7080, 8600/8700, MS2000/8000/9000	MaxFRAD, InterFRAD, HostFRAD, WEBrouter, EthrFRAD, MonFRAD, QuadFRAD	Marathon, NetRunner (500ET 75E, 1000E, NRI-R1000E)
Modular	Yes	So-so	Yes	Some models	Yes	Box level	Box level
Compact	Yes	Yes	Yes	Yes	Yes	Yes	Yes
Low cost	Yes	Not always	Not always	Not always	Yes	So-so	Yes
Support of embedded protocols (eg, Ethernet, SNA, and frame relay)	Yes	Yes, especially in big models	Poor support for SNA; no token ring	Poor support for SNA	Limited SNA	Yes	Yes
Easy to install, set up and maintain	Yes	So-so	Yes	Yes	Yes	Yes	Yes
Concentration of multimedia flows into a cost-effective uplink	Yes	No	Yes	Recent—no video	Partially	No, LAN focused	Yes
Ability to be integrated into enterprise/intranet environment	Yes	Yes	Yes	Yes	Yes	Yes	Yes
Support of data via protocols such as X.25, IP, SNA, and DDS	Yes	Yes, especially in big models	Yes	Yes	Subset	Yes	Yes
Support of voice using state-of-the-art compression schemes	Yes	No	Yes	Recent	Yes	No	Yes

TABLE 11.9 Typical Integrated FRAD Features (Continued)

	Momotec Products	Cisco Products	Hypercom Products	Motorola Products	ACT Products	Fastcomm Products	Micom Products
Support of video-conferencing over ISDN BRI connections	Yes	Some	Yes	Future	No	No	Yes
SNMP manageability	Yes	Yes	Yes	Yes	Yes	Yes	Yes
Routing capabilities for IP/IPX	Yes	Yes	Yes	Yes	Yes	Yes	Yes
Support of state-of-the-art routing algorithms (RIP2, OSPF)	Yes	Yes	Yes	Yes	Limited	RIP	Yes
Bridging capabilities (source routing and spanning tree)	Yes	Yes, especially in big models	Yes	Yes	Apparently not	Routing	Yes
Slot support	Medium (6)	Small in some models (1–2); more in other models (8–16)	Medium on LAN side (1–16); medium on WAN side (1–8)	Depends on model/price	Depends on model: 5–9 (Netperf), 1 Ethernet, 1 token, 4–8 voice	Depends on model: 1 serial/1 net, 4 serial/1 net, 1 Ethernet/3 serial/1 net	Depends on model: WAN 1–6, LAN: 1–12
Price range	$1600–4700	$1000–5000; much more in big models	$1750–14,500	$800–27,500	$2000–3000 (SDM); $25,000–40,000 (MS-2000)	$700 WEBrouter; $1000 Inter-FRAD to $14,000 Host-FRAD	$2500 (NetRunner 75E) to $12,000 (NRI-R1000)
Complete product line across enterprise	No	Yes	No	No, but extensive	No	No	No, but extensive

TABLE 11.9 Typical Integrated FRAD Features (Continued)

	Momotec Products	Cisco Products	Hypercom Products	Motorola Products	ACT Products	Fastcomm Products	Micom Products
Supports a cohesive networking philosophy (particularly routing)	Yes	Yes	Yes	Yes	Yes	Yes	Yes
Provides an integrated network management system	Partially via SNMP	Yes	Partially via SNMP	Partially via SNMP	Partially via SNMP	Partially via SNMP	Partially via SNMP
Vendor has the account and the corporate contacts	No	Yes	No	Some recognition	No	No	Some recognition

361

11.4 References

[act] Act Networks, Promotional Material, http://www.acti.com/vofr.htm.

[dmdp1] D. Minoli, *Voice over ATM,* DataPro Report, McGraw-Hill, June 1997.

[dmdp2] D. Minoli, *Voice over Frame Relay,* DataPro Report, McGraw-Hill, July 1997.

[em] E. Morgan, "Voice over ATM," ExpoComm'97 Conference Materials, Atlanta, June 1997.

INDEX

About the Authors

John Amoss, Ph.D., is a member of the technical staff at Lucent Technologies, where he is responsible for developing Lucent's data product strategy. Prior to joining Lucent he spent more than 20 years in broadband communications at Bell Labs and Bellcore. Amoss is also an Adjunct Professor of Telecommunications at Stevens Institute of Technology.

Daniel Minoli is Director of Engineering and Development, Broadband Services and Internet, at the Teleport Communications Group. A former developer of broadband communications technology at Bellcore, he also teaches as Adjunct Professor at Stevens Institute of Technology and New York University. Minoli has written more than 25 books on broadband communications.